Climate Change Management

Series editor

Walter Leal Filho, Hamburg, Germany

More information about this series at http://www.springer.com/series/8740

Walter Leal Filho · Anthony O. Esilaba
Karuturi P.C. Rao · Gummadi Sridhar
Editors

Adapting African Agriculture to Climate Change

Transforming Rural Livelihoods

The contribution of land and water management to enhanced food security and climate change adaptation and mitigation in the African continent

Springer

Editors
Walter Leal Filho
Faculty of Life Sciences
Hamburg University of Applied Sciences
Hamburg
Germany

Karuturi P.C. Rao
International Crops Research Institute for Semi-Arid Tropics
Addis Ababa
Ethiopia

Anthony O. Esilaba
Kenya Agricultural Research Institute
Nairobi
Kenya

Gummadi Sridhar
International Crops Research Institute for Semi-Arid Tropics
Addis Ababa
Ethiopia

ISSN 1610-2010 ISSN 1610-2002 (electronic)
Climate Change Management
ISBN 978-3-319-38592-1 ISBN 978-3-319-13000-2 (eBook)
DOI 10.1007/978-3-319-13000-2

Springer Cham Heidelberg New York Dordrecht London

Softcover reprint of the hardcover 1st edition 2015

Printed on acid-free paper

Springer International Publishing AG Switzerland is part of Springer Science+Business Media (www.springer.com)

Preface

Africa is one of the continents mostly severely affected by climate change, for two main reasons. The first reason is because the geographical characteristics of the African continent make it highly vulnerable to the effects of climate change, especially from the projected changes in the rainy seasons and intensitivity of droughts, which in turn may affect agriculture and other human activities.

The second reason for high vulnerability of African countries is related to their limited capacity to adapt. By not having access to required technological and financial resources that are needed to implement substantial adaptation programmes, many African nations are finding it difficult to handle the many challenges that climate change poses to them.

Climate change is also one of the major challenges that the agricultural research community is facing in recent years. Compared to many other biophysical constraints that the smallholder farmer is facing, climate change is a difficult problem to address for various reasons. First, climate change is a future problem and there are problems in assessing the magnitude and direction of these changes accurately, especially at local level. Second, while temperature projections seem to be fairly certain, changes in rainfall both in quantity and in variability are difficult to predict and rainfall is the major factor influencing productivity and profitability of the agricultural systems. Third, our understanding of impacts of projected changes in climate on crop growth and performance, especially the role of changes in carbon dioxide concentration, is limited. Despite these limitations, significant progress has been made in understanding the impacts of climate change on smallholder agricultural systems and in identifying appropriate management options to adapt. Unfortunately, much of the fieldwork carried out in many African countries remained inaccessible to the global community.

The conference "Transforming Rural Livelihoods in Africa: How can land and water management contribute to enhanced food security and address climate change adaptation and mitigation?" organized by the Soil Science Society of East Africa (SSSEA) in collaboration with African Soil Science Society (ASSS) and held in Nakuru, Kenya during 20–25 October 2013 served as an important platform for scientists in the Eastern Africa region to share their findings and experiences.

The targeted are researchers, policy makers, farmers, extension agents, among others, involved in and/or having interest in soil science and land and water management. This book contains various papers presented at the 2013 Nakuru Conference, as well as other contributions written by teams of African experts and/or by international researchers working in Africa.

Presentations at the conference covered a wide range of topics and presented a diverse set of viewpoints and perceptions on several of aspects of climate change and its impacts on agriculture. This book includes selected papers, based on their relevance and interest for the climate change research community, from the large number of presentations made during the conference. The papers are sequenced according to their focus in addressing a range of issues from methodological to technological and policy options for adapting agriculture to projected changes in climate.

Progressive changes in climate are hard to predict and assessing impacts of these changes on performance and productivity of crops is still harder. Since crop performance is an outcome of a number of interrelated factors it is difficult to predict how these factors independently and interactively affect the performance of crops under different climatic conditions. One of the promising approaches is the use of analogue sites, which are locations whose climate today appears as a likely analogue to the projected future climate of another location. The paper by Leal Filho and De Trincheria outlines this approach. The overall aim of climate change research is to find options that contribute to reduced vulnerability to climate variability and promotion of climate resilience in development investments, enhancing biodiversity, increasing yields and lowering greenhouse gas emissions. The second paper by Stephen Kimani highlights some of the measures that can be put in place to improve incomes and livelihoods of farmers in the semi-arid regions of Africa. The paper by Kwena Kizito dwells on the issue of how research generated information is availed and used. Through a review, this paper assessed the extent to which scientific information has been used to inform climate change adaptation policies, plans and strategies in Kenya as well as the effectiveness of existing platforms for sharing climate change information in the country. The paper by Sospeter Nyamwaro is based on information about the climate change-related projects undertaken in Kenya over the past five years. It analyses the areas covered by these projects and identified the high and low focus areas.

The next four papers deal with issues related to assessing and characterizing climate variability (Oscar Kisaka) and the potential impacts of climate variability and change on water resources (Sridhar Gummadi) and crop performance (Justice Nyamangara). One key aspect of climate change impact assessment studies is lack of information on how these impacts are felt differently by different gender, age and social class differentiated groups. The paper by Kumbiari Musiyiwa using the data collected through surveys conducted at analogue locations highlights this aspect of climate change and identifies gender sensitive adaptation options.

Among the key options for adapting agriculture to climate change, soil and water management measures including irrigation figure prominently. This is mainly because of the expected increase in the demand for water by crops due to increased evaporation and transpiration under warmer temperatures. The papers by Musyimi, Ngugi, Evans Mutuma and Geofrey Gathyungu provide some insights into the potential role of water conservation in mitigating the water stress on some important food crops. The study reported by P.N.M. Njeru tried to compare and contrast farmer and scientific evaluation of various climate change adaptation options that integrate soil water and soil fertility management practices aimed at improving productivity of sorghum.

The final set of four papers explores the use of drought tolerant crops and varieties as an alternative adaptation strategy. Finyange N. Pole evaluated a number of maize genotypes to identify varieties that are efficient in both nutrient and water use. While Fabian Bagarama explored the performance of tomato as an alternate crop under warmer climates, studies reported by Cyrus M. Githunguri assessed the potential of traditional food crops as alternatives. Interest in research on issues related to climate change in Africa has been high over the past decade. It is important that this remains high and these efforts will be successful in identifying robust management options that help smallholder farmers make best use of the variable climatic conditions while helping in adapting to future changes.

This book is also an output of the project Adapting agriculture to climate change: Developing promising strategies using analogue locations in Eastern and Southern Africa (CALESA), funded by the German International Agency for Cooperation (GIZ) and undertaken by the International Crops Research Institute for the Semi-Arid Tropics (ICRISAT) in collaboration with Kenya Agricultural Research Institute (KARI), Kenya Meteorological Department (KMD), Zimbabwe Meteorological Department (ZMD), Midlands State University (MSU) and the Hamburg University of Applied Sciences (HAW) in Germany.

Using a combination of model-based ex ante analyses and iterative field-based research on station and in farmers' fields, the project has tested potential agricultural adaptation strategies for rainfed agriculture in the semi-arid and dry sub-humid tropics. This has been achieved through choosing four currently important crop production zones (two in Kenya and two in Zimbabwe) and then identifying corresponding 'spatial analogue locations' for each production zone, providing eight study locations in all. This book contains a set of chapters which describe some of the results achieved as part of the project.

The editors wish to thank the GIZ, the CALESA project partners, the Soil Science Society of East Africa (SSSEA) and the Africa Soil Science Society (ASSS), for their support to the conference, to the CALESA project and to this book. The ASSS and the SSSEA acknowledge, with appreciation, the efforts and contributions of the Kenyan government, Kenya Agricultural Research Institute (KARI), ICRISAT, the Alliance for a Green Revolution in Africa (AGRA), the National Commission for Science Technology and Innovations (NACOSTI), MEA Ltd, The International Atomic Energy Agency (IAEA), The Association for Strengthening Agricultural Research in East and Central Africa (ASARECA), Africa Soil Health Consortium (ASHC), the International Union of Soil Science (IUSS), SANREM Innovation Laboratory of Virginia Tech, Australian Agency for

International Development (AusAID), the University of Sydney and the Joint Research Commission (JRC) of the European Union (EU) for supporting the conference.

Due to its scope, the actuality of the topic and its importance in documenting and promoting experiences of climate change adaptation in Africa, this book will provide timely assistance to the current and future adaptation efforts in the African continent.

Walter Leal Filho
Anthony O. Esilaba
Karuturi P.C. Rao
Gummadi Sridhar

Contents

Chapter 1
Adapting Agriculture to Climate Change by Developing Promising Strategies Using Analogue Locations in Eastern and Southern Africa: A Systematic Approach to Develop Practical Solutions

J. De Trincheria, P. Craufurd, D. Harris, F. Mannke, J. Nyamangara, K.P.C. Rao and W. Leal Filho

Abstract From 2011 to 2014, the CALESA project was a research-for-development project which coupled integrated climate risk analyses, crop growth simulation modelling and field-based research both on-station and on-the-ground with participatory research with farmers. It comprised research-oriented activities for knowledge and technology creation, and development-oriented activities for information sharing and capacity building. The main purpose of the CALESA project was to develop sound adaptation strategies for future temperature increases associated with greenhouse gas emissions using "analogue locations", both as learning- and technology-testing sites. This was meant to improve the ability of rainfed farmers in the semi-arid tropics of sub-Saharan Africa, in particular Kenya and Zimbabwe, to adapt to progressive climate change through crop, soil and water management innovation, and appropriate crop genotype choices. Another key feature of the CALESA project was the development and implementation of tailor-made capacity-building activities specifically designed to fulfil the needs of local scientists in the field of climate change adaptation and climate-smart agriculture. To achieve its objectives, the CALESA project used a combination of model-based ex ante analyses and iterative field-based research on station and in farmers' fields. This facilitated the evaluation of potential agricultural adaptation strategies for

J. De Trincheria (✉) · F. Mannke · W. Leal Filho
Faculty of Life Sciences, Hamburg University of Applied Sciences, Ulmienlet 20, 21033 Hamburg, Germany
e-mail: JosepMaria.DeTrincheriaGomez@haw-hamburg.de

P. Craufurd · D. Harris · K.P.C. Rao
International Crops Research Institute for the Semi-Arid Tropics (ICRISAT), P.O. Box 29369, Nairobi, Kenya

J. Nyamangara
International Crops Research Institute for the Semi-Arid Tropics (ICRISAT), Bulawayo, Zimbabwe

W. Leal Filho et al. (eds.), *Adapting African Agriculture to Climate Change*, Climate Change Management, DOI 10.1007/978-3-319-13000-2_1

rainfed agriculture in the semi-arid and dry sub-humid tropics. In this line, four important crop production zones (two in Kenya and two in Zimbabwe) were identified. Subsequently, the corresponding 'spatial analogue locations' for each production zone, providing eight study locations in all, were identified. A strong element of participatory research with small-scale farmers ensured that the perceptions of current and future climate risk and their preferred climate change adaptation strategies was effectively taken into account. In addition, this also ensured that the project activities and outputs remained relevant to their needs and expectations. The main outputs of the CALESA project are as it follows. Firstly, the identification and fully characterisation of four important crop growing areas in Kenya and Zimbabwe which comprise cool/dry, cool/wet, warm/dry and warm/wet growing conditions, and their temperature analogue locations. Secondly, through the combined use of long-term daily climate data, crop growth simulation models and participatory surveys with farmers, the identification and quantification of the implications of both current and future climate change production risk at the study locations. Thirdly, through iterative field research both on station and in farmers' fields over more than 2 years, the evaluation of potential crop, soil and water management, and crop genotype adaptation options. This was followed by the formulation of adaptation strategies for the target locations. Finally, through the overall implementation of the project activities, the institutional capacity in understanding climate change impacts and the development of effective adaptation responses in Kenya and Zimbabwe was fostered.

Keywords Rainfed agriculture · Climate change adaptation · Temperature analogue locations · Climate modelling · Eastern and southern Africa

Introduction

Between now and 2050, the world's population will increase by one-third, mostly in the developing world. Today, there are still 1,870 million people estimated to have severe nourishment deficiencies in the world (FAO 2012). In addition, FAO state that there are 20 countries in sub-Saharan Africa (SSA) which periodically face food crises.

In most of these countries, paradoxically, agriculture is an important, if not the major, part of economy. In this line, rainfed agriculture is vital and expected to remain like this in order to ensure food security in SSA. Nearly 90 % of staple food production will continue to come from rainfed farming systems (Rosegrant et al. 2002). However, there are special challenges to the development of SSA's rain-fed farming systems. On one hand, it is in SSA where some of the poorest and most vulnerable communities live: 40 % of the continent's population lives with less of USD 1 day^{-1} and 70 % of these communities are in rural areas (Chen and Ravallion 2007). On the other hand, rainfed agriculture has stagnated. Furthermore, in addition to the constraints imposed by policy failures, extreme poverty and often a degrading resource base (Sanchez 2002), the inherent climate-induced production risk associated with the current high level of season-to-season spatial and temporal variability of rainfall in the semi-arid and dry sub-humid tropics also acts as a key challenge (Christensen et al. 2007). Impoverished farmers are risk averse and are unwilling to invest their assets at hand in costly innovations when the outcomes seem so uncertain from season to season (Cooper et al. 2009, 2008).

It seems clear now that there is not any reasonable doubt about the link between human activity and global warming (IPCC 2007). IPCC (2007) also states that rainfed agriculture is likely to be worsened by global warming and its predicted impacts on seasonal rainfall amounts and distribution patterns. This threatens to exacerbate the climate-induced risk problems already faced by rainfed farmers. However, predicting the exact rate, nature and magnitude of changes in temperature and rainfall is a complex scientific undertaking. Especially, there is currently considerable uncertainty with regard to the final outcome of climate change and its impacts. Whilst such predictions continue to remain uncertain, most key investors in agricultural development in low-income economies agree that it is the poor and vulnerable who will be the most susceptible to changes in climate (DFID 2005). This is particularly true for those communities in SSA who rely on rainfed agriculture and/or pastoralism for their livelihoods. Such communities, already struggling to cope effectively with the impacts of current rainfall variability, will face major problems to effectively adapt to future climate change.

Even though all General Circulation Models (GCM's) models agree that it will become warmer across sub-Saharan Africa, the degree of warming predicted is quite variable. The Fourth Assessment Report of the IPCC suggests that the median temperature rise in eastern and southern Africa will be 3–4 °C by the end of the 21st century (Christensen et al. 2007). These authors also state that there will be a greater temperature rise in June, July and August (median rise 3.4 °C) than from September to February (3.1 °C) in eastern Africa and a greater temperature rise in September, October and November (3.7 °C) than in December to May (3.1 °C) in southern Africa. Indeed, evidence of changes in climate extremes, in particular with regard to temperature, is already emerging in southern and West Africa (New et al. 2006). However, with regard to the percentage of changes in rainfall amounts, the uncertainty is considerably greater. Nevertheless, there appears to be a consensus predicted trend of wetting in eastern Africa and of drying in the winter rainfall regions of southern Africa. Christensen et al. (2007) predicted a median increase in rainfall of 7 % (2–11 % for the 25th and 75th percentiles) for eastern Africa, while annual rainfall is predicted to decrease by 4 % (−9 to +2 % for the 25th and 75th percentiles) for southern Africa by the end of the 21st century.

Current climate variability and future climate change may lead to reduced availability and access to natural resources, as well as a diminishing of the livelihood and welfare for rural communities in SSA (Galloway 2010). In addition, even though these communities have contributed very little to GHG emissions, they are expected to be the group most severely affected by its impacts (Reid et al. 2010).

This high vulnerability to future climate change is mainly caused by a low adaptive capacity to external stresses and changes in their environment on one hand, a direct dependence on sensitive sectors and natural resources for their livelihood and sustenance on the other. In addition, this situation is worsened by marginally available financial resources and know-how for designing and implementing effective adaptation measures (Galloway 2010; Reid et al. 2010).

Climate change has already significantly impacted agriculture and it is expected to further impact directly and indirectly food production. This report also states that the extent of these impacts will depend not only on the intensity and timing (periodicity) of the changes but also on their combination, which are more uncertain, and on local conditions. Therefore, anticipating appropriately the impacts of climate change on agriculture requires data, tools and models at the spatial scale of actual production areas.

FAO states that the impacts of climate change will have major effects on agricultural production, with a decrease of production in certain areas and increased variability of production in other areas. Among the most affected areas are economically vulnerable countries already food insecure and some important food exporting countries. Consequently, climate change is expected to increase the gap between developed and developing countries as a result of more severe impacts in already vulnerable developing regions, exacerbated by their relatively lower technical and economical capacity to respond to new threats (IPCC 2007).

Agriculture has to address simultaneously three intertwined challenges: ensuring food security through increased productivity and income, adapting to climate change and contributing to climate change mitigation (FAO 2010). FAO states to contribute addressing these three intertwined challenges, food systems have to become, at the same time and at every scale from the farm to the global level, more efficient and resilient. Food systems have to become more efficient in resource use and become more resilient to changes and shocks.

Given the constraint of both current climate induced production risk and the predicted change in nature of that risk in the future, it seems now evident that a two pronged approach to adaptation to climate change is required (Burton and van Aalst 2004; DFID 2005; Washington et al. 2006; Cooper et al. 2008; ICRISAT 2008). In the short to medium term, it is essential to help poor and vulnerable farmers to build their livelihood resilience (and hence adaptive capacity). This has to be made through coping better with current climate-induced risk as a pre-requisite to adapting to future climate change. In the medium to longer term, and as climate change begins to bite, farmers will have to progressively adapt their farming practices to a new set of climate induced risks and opportunities. It is on the second 'medium to longer term' aspect of adaptation that the CALESA project has placed its emphasis, specifically, effective adaptation to progressive climate change.

Goals and Objectives of the Project

The main goal of the project "Developing promising strategies using analogue locations in Eastern and Southern Africa (CALESA)" was to improve the ability of rainfed farmers in the semi-arid tropics of Africa to adapt to progressive climate change through crop, soil and water management innovations, and appropriate crop genotype choices.

To achieve this goal, the project developed sound adaptation strategies for future temperature increases associated with greenhouse gas emissions using 'analogue locations', both as learning sites and as technology testing sites. In addition, the CALESA project combined simulations and field assessments, with an analysis of the views and perceptions of relevant stakeholders, paying a special attention to gender issues.

The CALESA project has been funded by the German Agency for International Cooperation (*Deutsche Gesellschaft für Internationale Zusammenarbeit, GIZ*) on behalf of the German Ministry of Cooperation and Development (*Bundesministerium für wirtschaftliche Zusammenarbeit und Entwicklung, BMZ*). The project was also supported by the International Climate Change Information Programme (ICCIP), which assisted with the dissemination elements.

The project was coordinated by the International Crop Research Institute for the Semi-Arid Tropics (ICRISAT). The cooperation partners of the CALESA project were:

- Kenya Meteorological Department (KMD), Kenya.
- Kenya Agricultural Research Institute (KARI), Kenya
- Midlands State University (MSU), Zimbabwe
- Zimbabwe Meteorological Department (ZMD), Zimbabwe
- Hamburg University of Applied Sciences, Faculty of Life Sciences, Germany

Methodology

The CALESA project used a combination of model-based ex ante analyses coupled with iterative field-based research on station and in farmers' fields. This was meant to test potential agricultural adaptation strategies for rainfed agriculture in the semi-arid and dry sub-humid tropics. This was achieved by means of choosing four currently important crop production zones (two in Kenya and two in Zimbabwe) and then identifying corresponding 'spatial analogue locations' for each production zone, providing eight study locations in all.

The project defined "analogue locations" as those locations that have *today* the climatic characteristics that are expected *tomorrow*. In defining the locations, special attention was given to adaptation to temperature increases. Altitudinal effects on mean air temperature facilitated this. Given the potential of 'analogue locations' to provide a solid basis for such research across sub-Saharan Africa, special attention was also given to the continuous documentation and dissemination of project activities and achievements through the web, newsletters and dissemination events. A strong element of participatory research with farmers within the project locations ensured that the project activities and outputs were relevant to their needs and expectations.

In order to achieve the goals and objectives of this research, the project evaluated agricultural adaptation strategies to climate change by means of the use of 'analogue locations'. In this line, a special focus was exerted on predicted temperature increases. A clear relationship between altitude, air temperatures and crop growth and yield facilitated the identification of appropriate temperature analogue locations in eastern and southern Africa. Across these two countries the project identified analogue locations for four food production areas that currently experience cool/dry, cool/wet, warm/dry and warm/wet growing conditions.

To sum up, the following areas of research were addressed:

1. Development of criteria for the selection of analogue locations and identification of four paired analogue locations for four currently important food production areas across Kenya and Zimbabwe using CLIMEX software (www.climatemodel.com/climex.htm).
2. Access to long-term daily climatic data for those paired locations (40 years +) and development of detailed climate risk and climate trend analyses through the use of the statistical package In-Stat (www.graphpad.com/instat/instat.htm)
3. Full characterisation of the four paired sets of locations with regard to crops, soils, climate, current farming practices, the roles of men and women, crop diversity, livestock management, farmers' perceptions of current climate-induced risk and climate change, and possible adaptation strategies. This was done through participatory research with farming communities. Added depth was given to this by means of two PhD students which focused their PhD research on the related gender aspects of the variables previously mentioned.
4. Field calibration of the weather-driven Agricultural Production Systems Simulator (APSIM) for important locally grown food crops at the four sets of paired locations (http://www.apsim.info/apsim/Documentation/). This was done through detailed on-station agronomic and physiological research.
5. Use of GIS, downscaled GCM predictions and other innovative tools for incorporating and extrapolating spatial and temporal effects of climate change, including gender and environmental impacts.
6. Iteratively testing of the potential of improved soil, water and crop management strategies together with contrasting crop genotypes to mitigate the impacts of increased temperature. This was done by means of a combination of field research on station and in farmers' fields and simulation based research over the 3-year period.
7. Special attention was given to informing and communicating the project outputs through the use of the web, logo and poster development, newsletters, and dissemination events.

Stakeholders

Stakeholders of the CALESA project have included small-scale farmers and other farmer groups within the chosen production areas and their analogue locations in Zimbabwe and Kenya. Furthermore, staff of the national agricultural systems (NARS) and national meteorological services (NMS) in these two countries received hands-on training on climate risk analyses, participatory interaction with farmers and the approaches associated with the use of analogue locations. The information gained through testing the use of analogue locations as a tool to evaluate adaptation strategies with farmers is expected to potentially be of great value to NARS and NMS, not only in Kenya and Zimbabwe, but in all countries in SSA where rainfed agriculture is important. Furthermore, 2 postgraduate students from Kenya and Zimbabwe, which have been supervised by Hamburg University of Applied Sciences and Manchester Metropolitan University, have gained extensive experience of evaluating the gender related aspects of agricultural climate adaptation strategies through the use of analogue locations.

Immediate beneficiaries have been the NMS and NARS staff based at the eight chosen locations whose skills in climate analyses and adaptation to climate change science has been strengthened. Importantly, they have experienced the benefits of agricultural and meteorological collaboration in the service of supporting rainfed farming systems. In the longer term, it is expected that the lessons learned with regard to the use of analogue locations as a tool to evaluate climate change adaptation strategies can benefit national policy makers, NARS and NMS, and farmers in all SSA countries. To ensure the latter, the project gave specific priority attention to the dissemination of its activities and results. Ultimately, it is small-scale rainfed farmers who will be in a better position to make use of the results of climate change adaptation research and be able to ensure their future livelihoods in a warming world.

Outputs

A critical analysis of the results and drawing key lessons of the CALESA project is presented as chapters of this book. In addition, several papers have been published in international peer-reviewed scientific journals during the lifetime of the project. In this line, prospective papers outlining key transnational results and lessons learnt are also expected to be published in a continuous basis during 2014. Therefore, this section aims to present a summary of the main outputs of the CALESA project, and the rationale behind of them.

Identification and Biophysical Characterisation of the Temperature Analogue Locations in Kenya and Zimbabwe

Identification of the Temperature Analogue Locations

In the frame of the CALESA project, four currently important crop growing areas in Kenya and Zimbabwe which comprise cool/dry, cool/wet, warm/dry and warm/wet growing conditions, and their temperature analogue locations, have been identified and fully characterized.

Much of the past work on assessing the agricultural impacts of climate change is based on regional climate change scenarios developed from GCM outputs. One major problem with this approach is its inability to capture the variability and associated impacts at local scales which are essential for planning and development of adaptation strategies. An alternative approach is to use climate analogues which can serve as plausible descriptions of possible future climate. Thus climate analogues can be used to assess impacts of warmer climates on crop production, understand the main features of adaptation and test alternative management options that help mitigate the negative impacts of climate change. In this line, the CALESA project has used spatial analogues in order to characterise and understand the impacts of climate change on crops and cropping systems which are relevant to semi-arid and dry-sub humid tropics in eastern and southern Africa.

A key step towards the achievement of this output was the identification of analogue climates that mimic what might be expected under climate change. This was done using a carefully developed criterion which includes all parameters that affect crop production. Most important were maximum and minimum temperature, and amount and seasonality in rainfall. The criteria also considered the biophysical conditions, crops and cropping systems, and other social and economic drivers. This was meant to ensure that observed differences are directly linked to differences in climate. Tools like CLIMEX and GIS were used to locate the analogues by matching the variables selected (Table 1.1).

Table 1.1 Temperature analogue locations of the CALESA project in Kenya and Zimbabwe

Kenya	°C[a]	Zimbabwe	°C[a]
Embu KARI research station	19,5	Sanyati cotton research station, Kadoma	21,8
Kabete University of Nairobi farm	18,2	Chiweshe Henderson research station	18,2
Katumani KARI research station	19,2	Chiredzi Chiredzi research station	21,3
Kampi-Ya-Mawe KARI research station	20,8	Matobo Matopos research station	18,4
Ol Jororok KARI research station	14,9		

[a] Average annual temperature

Socio-economic Characterisation of the Temperature Analogue Locations

Focusing on the locations identified and working with NARS and NMS partners, necessary data on long-term climate, crop production, and socioeconomic characteristics was collected from secondary sources. This was meant to fully characterise the environments highlighting the similarities and dissimilarities on one hand, and any long-term trends in climate and crop production on the other. In addition to matching climate variables for amount and variability, special attention was paid to conditions such as day length, vegetation cover, soil type, proximity to water bodies, and other geographic features. All the data was spatially referenced for use with GIS.

In addition, in order to gain in-depth understanding into stakeholder perceptions of current climate variability and future climate change, their impacts on agricultural systems, and document indigenous knowledge, consultations with key stakeholders were carried out during the first year of the implantation of the project. This was done in close collaboration with NARS and NMS partners in Kenya and Zimbabwe. The project considered stakeholders at all levels, from farmers to policy makers. Stakeholder interest, knowledge, attitude and practices were solicited through stakeholders' workshops and semi-structured surveys.

As a key activity part of the socio-economic characterisation of the selected locations, the CALESA project undertook a set of baseline surveys to explore the socio-economic environment of the pairs of analogue locations in Kenya and Zimbabwe. Through participatory research with farming communities, the farmers' perceptions with regard to current climate variability and future climate change, and their impacts on the agricultural systems in each of the locations were assessed. The purpose of the survey was to characterize smallholder agricultural practices at all the study sites. Specifically, the socio-economic characteristics of household at reference sites in comparison with those at analogue sites; crop diversity and management practices between analogue sets and across analogue sites; livestock management strategies between analogue sets and across analogue sites; and constraints to agricultural production at reference sites and analogue sites. This was meant to identify and quantify production risks at the wetter and drier analogue pairs for 2050s climate. A total of 722 respondents in Kenya and 627 respondents in Zimbabwe were interviewed using structured questionnaires. The data from semi-structured questionnaires was analysed using SPSS statistical package. The survey results were presented to the stakeholders for feedback and refinement (Fig. 1.1).

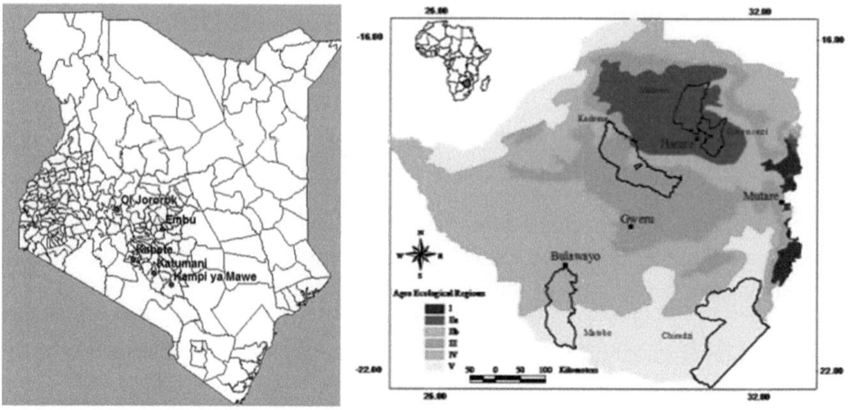

Fig. 1.1 Temperature analogue locations of the CALESA project in Kenya and Zimbabwe

Quantification of the Performance of Crop, Soil and Water Management, and Crop Genotype Adaptation Options

While the demand for systematic practices and technologies for adaptation to climate change is growing, information required to formulate such strategies is not available. This is especially the case with regard to how crops, varieties and management practices perform under different hydrological and thermal regimes under different farmer managed conditions. Thus the work performed aimed to develop this information through a series of well-planned field trials which facilitated the assessment of the biophysical performance, profitability, feasibility and end user acceptability of potential adaptation options under realistic farmer conditions. This was done using a 'mother-baby' trial approach and through a critical assessment of the results from the field trials. The assessment was enhanced over extended time periods through the use of APSIM and downscaled GCM climate change predictions.

Three main activities were performed to achieve this aim. Firstly, field trials and collection of detailed multi-dimensional data. This was required to facilitate the development of a critical assessment of the performance of management options at selected locations. Secondly, the adequacy of management options to cope with the predicted changes in climate was evaluated. Finally, climate change adaptation strategies for target locations were formulated.

In this line, crops and agricultural management practices were identified and selected for further evaluation. Thus, a set of four field trials was initiated at selected climate analogue sites in Kenya and Zimbabwe. This was meant to assess and quantify the performance of various crops and management practices under different temperature regimes. By means of these carefully planned trials, the CALESA project aimed to generate primary data on how temperature affects crop growth and development, and how management can be adapted to mitigate the stresses associated with increased temperatures. The data collected served to calibrate and validate the crop simulation model APSIM. This was meant to generate a more realistic regional assessment of impacts of climate change on agriculture in Kenya and Zimbabwe. Furthermore, best bet options, which may facilitate adaptation to current and future variability in the climatic conditions, were identified as well. In addition, detailed measurements on phenology and crop growth were made to quantify the crop and varietal response to changes in temperature.

The trials can be outlined as it follows:

Trial 1: Crops and Varieties: The purpose of this trial was to assess the performance of crops (maize, sorghum, groundnut and cowpea) under different climatic regimes. Different varieties of these crops were also included in the assessment.

Trial 2: Moisture conservation and plant population: This trial was designed to determine the effect of water conservation and plant population on productivity of legumes (groundnut and cowpea). Medium maturity varieties of groundnut (variety Nyanda) and cowpea (CBC 2) were planted in the trial. Grain yields were significantly different between the dry sites at Matopos (cool/dry) and Chiredzi (hot/dry) research stations whereas groundnut yields were significantly different between wet sites at Kadoma (hot/wet) and Mazowe (cool/wet) research stations.

Trial 3: Moisture conservation and fertility: This trial sought to determine the effects of water conservation (tillage) and fertilizer application on the productivity of maize and sorghum.

Trial 4: Adjustments to planting dates and planting methods: The objective of this trial was to evaluate the effect of seed treatment and planting dates on the performance of maize, sorghum, groundnut and cowpea. The seed treatments were dry seed, priming, and priming + GroPlus (GroPlus is a phosphorus-rich soluble starter fertilizer applied to primed seed). Late planting generally gave low yields for all crops at each study site compared with early planting.

Preferred agricultural adaptation options of farmers (differentiating between men and women preferences) under changing climates: In this line, farmers participated in evaluating on-station trials of different crop adaptation options during the implementation of the field trials. This was meant to capacitate and empower farmers with adequate knowledge on efficient climate-smart agricultural practices for current and future climates.

Ex ante assessment to quantify the current climate risks and yield gap and risks: This was carried out with the crop simulation model APSIM using readily available data on one hand and synthetic scenarios ('arbitrary' or 'incremental' scenarios) and GCM derived scenarios on the other. The results were subjected to statistical and economic analysis as well as stakeholder evaluation. Based on the outcome, the management options were classified into low impact, medium impact and high impact groups in relation to their climate sensitivity and climate-induced risk.

Implications of Both Current and Future Climate Change Production Risks at the Study Locations in Kenya and Zimbabwe

System models such DSSAT and APSIM include simulation of temperature effects on crop growth and development processes, and in some crops, heat stress effects on seed development. A key insight from recent climate change analysis using APSIM revolves around the fact that the deployment of existing longer duration cultivars could be a first level response in adapting to future climate change. This project tested this response strategy, along with other key crop and soil management options, by monitoring the growth and development of crop cultivars across a range of sites which represented the temperature increases predicted under future climate change. At the same time, the field studies provided an opportunity to further evaluate and improve the simulation models to capture temperature effects on plant growth. Of particular interest was the question of whether extreme high temperatures effects on plant growth processes were adequately described in the models.

The main activities carried out were in the first place to enhance the NMS and NARS capacity to build, analyse and utilize high quality climate data. In addition, the performance of the APSIM in simulating temperature effects of growth and yield of important food crops was evaluated. Furthermore, the field studies were used to calibrate cultivar parameters and validate APSIM simulation of temperature effects.

Crop simulation models are the key tools to assess the net impact of climate change on agriculture. These models integrate scientific knowledge from many different disciplines (crop physiology, agronomy, agrometeorology, soil water, etc.) and help in holistic assessment of performance of crops and systems under different soil, climatic and management conditions. In addition, they are increasingly used to understand the effects of climate variability and change.

In this line, MarkSim-GCM, which is a tool developed to generate location-specific weather data for future climates, was used to develop baseline and future climates to mid and end century periods for the five CALESA project locations in Kenya. In addition, climate data from four locations in Kenya was analysed for variability and trends in rainfall and temperature for the period 1980–2010. The uncertainty and risk associated with rainfall is one of the major factors influencing farmers to adopt low input technologies that are low in risk but also low in productivity.

An ex ante analysis was carried out to assess the climate sensitivity of these technologies using observed and downscaled location specific climate scenarios for one site location in Kenya. Crop simulation model APSIM 7.4 was used to generate yields of maize for 30 years using observed weather data and up-scaled weather data for 6 GCMs (BCCR, CNRM, ECHAM, INMNM, MIROC and CSIRO) as well as the average (ensemble) for A2 carbon emission scenarios during the end century (2070–2099). Role of plant population, planting time, variety, application of fertilizer, and use of soil and water conservation technologies in adapting to climate change were considered in this analysis.

Strengthening of the institutional capacity in understanding climate change impacts and developing effective adaptation responses.

These activities included aspects of information, education, communication and training on one hand, and effective dissemination of the project results on the other. This entailed the coordination of the logistics, organisation and monitoring of the training activities, evaluation of the outputs and outcomes of the project activities, and the provision of any relevant recommendations. The work also entailed the formulation of strategies for the promotion and dissemination of the project and its activities, as well as the long-term sustainability of the project outputs.

Hands-on Training of Local Scientists and Local Communities

Given the general complexity and extreme variability associated with climate parameters, it is difficult to characterise and understand long-term trends in the climate as well as their impacts on agricultural systems. Analysis of such highly variable data requires advanced tools and methods that systematically look through large amounts of scattered data for trends, and assess its agricultural consequences.

In recent years, aided by the rapid advances in computing technology, several science-based models and approaches were developed to analyse and summarise climate information, assess its potential impacts on agricultural systems, and conduct scenario-analyses. These were meant to aid the identification of promising adaptation strategies to climate variability and change. However, the use of these tools by researchers from most African countries remained very low. This was mainly due to lack of skills and experience in using them, and also due to lack of awareness about their potential application.

Over the past three years, the CALESA project developed and implemented several hands-on training programs which were meant to enhance the capacity of the African project team members to analyse climate data and to assess climate impacts on the performance of agricultural systems.

Three key areas have been the target of the capacity-building activities of the CALESA project. These activities include the characterisation of the variability associated with current and future climatic conditions, the assessment of the impacts of climate variability on productivity and profitability of various crops and cropping systems, and an ex ante assessment of risks and opportunities created by variable climatic conditions to guide climate change adaptation planning.

The project used formal training as well as hands-on-work as means to enhance the capacity of partners in the use of selected tools. These capacity-building activities were mainly aimed at:

- Improving the skills of researchers in the analysis of long-term climate data and characterise variability and trends in climate;
- Introducing the stochastic climate models ("weather generators") to generate long-term climate data including future climates for use with crop simulation modcls;
- Improving application and understanding of crop simulation model APSIM to characterize and quantify climate impacts on agriculture;
- Hands-on-work planning and conducting of various trials, and collection of good quality data, as required to calibrate and validate crop simulation models (Fig. 1.2).

To achieve this, the CALESA project developed and/or identified required tools, and trained the partners in using them. The tools identified and used in the training and field work include the following:

- Simple spreadsheet based tools: "Temperature Analyser" and "Rainfall Analyser" for a quick assessment of variability and trends in temperature and rainfall;
- Stochastic weather generator MarkSim-GCM to generate required climate data to fill gaps in the observed data and generate downscaled location specific future climate scenarios;
- Crop simulation model APSIM (Agricultural Production Systems Simulator) to quantify climate impacts on productivity and sustainability of agricultural systems under current and projected climatic conditions;
- Protocols to measure various soil, plant physiological, and crop growth parameters, as it is required to calibrate and validate crop models.

Accordingly, a training module that included a set of tools ranging from simple spreadsheet based models to complex system simulation models was developed and implemented in Kenya.

Fig. 1.2 Hands-on capacity-building activities for local scientists in Kenya (*left*) and Zimbabwe (*right*)

Box 1 The tools and models covered by the training program

- Rainfall and temperature analysers: Spreadsheet based tools to analyse up to 50 years of rainfall and temperature data as well as to generate information that can help characterise temporal variability and trends in the monthly, decadal, weekly, seasonal and annual rainfall amounts at any given location.
- MARKSIM-GCM: A web-based stochastic daily weather generator based on a third order Markov chain model. It is a very useful tool to generate site-specific climatic data for locations where there is no observed data available or where it is incomplete. Such data is essential for running crop simulation models. It also includes an option of simulating location-specific future climatic conditions for different emission scenarios using downscaled GCM outputs.
- APSIM: A system simulation model with capabilities to simulate the growth and yield of a range of crops in response to changes in soil, climate and management practices under current and future climatic conditions. When calibrated and validated for local conditions, the model serves as a valuable tool to assess the impacts of climate variability and change on productivity, profitability and sustainability of the agricultural systems.
- Risk analyser: A spread sheet based tool to estimate the production costs of various crops and cropping systems and construct both risk and return profiles using long-term simulated production data.

The trained team members are expected to serve as master trainers for further training in their respective institutions and countries. The project conducted training programs on 2012 with participants from Kenya Agricultural Research Institute (KARI), Kenya Meteorological Department (KMD) and ICRISAT-Kenya. A second training program, which was aimed at improving the technical capabilities and skills of research technicians and scientists on planning and managing scientific trials, and on the collection of high quality data, was conducted later on. The scientists and research technicians involved in this training program were the ones currently responsible for managing the planned trials at the five selected locations in Kenya.

The subjects covered included systemic planning and conduction of multiple trials, use of equipment such as moisture meters and growth analysers, use of field note books for systemic collection and recording of data, and data analysis and archival methods. These formal trainings were followed by a number of follow-up activities to ensure that the trainees put the skills and knowledge acquired during the training into practice. The follow-up activities included on-line support to address any constraints, on-site support to resolve practical problems encountered with the use of methods and equipment, and a refresher course to review and share experiences and upgrade the skills.

All these activities have made significant contributions to enhance the skills of all the scientists and research technicians associated with the CALESA project in Kenya. In one of the evaluations conducted to assess the benefits of these capacity enhancement efforts, participants evaluated them as extremely useful giving them a score of 4.64 on a scale of 0–5. These efforts have led to the establishment of a core team of researchers with skills and experience to conduct advanced research on issues related to climate variability and change in Kenya. Participants of this training are now actively participating and contributing to various other projects including the Agricultural Model Intercomparison and Improvement Project (AgMIP), which is currently being implemented by their respective home institutions. Effective training, practical follow-up actions, a continuous engagement in project activities, and the availability of data required to work with advanced software-based tools, were the major contributors to the success of the capacity-building efforts under the CALESA project.

PhD Research in the Frame of the CALESA Project

Capacity-building within the framework of the CALESA project also involved two PhD studentships focusing on the gender aspects associated with the management of current and future agricultural climate-induced risk. The PhD research was carried out by two young African researchers from Kenya and Zimbabwe. The selection of the candidates was based on an open call for applications issued by Hamburg University of Applied Sciences (HAW Hamburg) via the International Climate Change Information Programme (ICCIP: http://www.iccip.net). Since African women are still underrepresented in respect of post-graduate qualifications in general and in climate research in particular, female candidates were particularly encouraged to apply. As a result of this selection process, the CALESA project finally selected two PhD candidates: Jokastah Kalungu (Kenya) and Kumbirai Musiyiwa (Zimbabwe).

The two PhD candidates were supervised by Professor Walter Leal (HAW Hamburg, Germany) with the collaboration of Dr. Dave Harris (ICRISAT-Kenya) and Dr. Justice Nyamangara (ICRISAT-Zimbabwe). The PhD candidate from Kenya focused her research on an assessment of impacts of climate change on smallholder farming practices and the role of gender on adaptation strategies in semi-arid and sub-humid regions of Kenya. The PhD candidate from Zimbabwe focused her research on assessing climate-induced risks and gender sensitive adaptation options in Zimbabwe.

In the frame of the CALESA project, the two PhD candidates have been engaged in all the activities across the selected locations in Kenya and Zimbabwe. Furthermore, cooperative research focusing on climate-smart agricultural management in sub-Saharan Africa was also conducted by the two PhD candidates, HAW Hamburg (Germany), and the ICRISAT teams in Kenya and Zimbabwe. Each of the PhD candidates had to develop at least three scientific papers which had to be submitted to international renowned peer-reviewed journals. In addition, the two PhD candidates have also attended several international conferences where they have presented their research.

As a key part of the PhD programme, the two PhD candidates annually attended several research capacity-building seminars in Hamburg (Germany). These tailor-made seminars were specifically designed to respond to the needs of the two PhD candidates and fulfil the requirements of high-quality PhD research in Europe. Invited speakers during the seminar were international renowned experts on the field of climate-change adaptation and agriculture, gender, and research capacity-building. The two PhD candidates submitted and defended their PhD dissertation on 2014 in Manchester (United Kingdom).

The Final CALESA Conference

The final workshop of the CALESA project took place on October 2013, in Nakuru (Kenya). It was organised as a special session under the 2013s Africa Soil Science Society (ASSS) and the Soil Science Society of East Africa (SSSEA) Joint International Conference. The theme of this conference, which was opened by the representatives of the Kenyan authorities, was: "Transforming Rural Livelihoods in Africa: How can land and water management contribute to enhanced food security and address climate change adaptation and mitigation?".

The conference attracted over 200 professionals and practitioners in agriculture and rural development, from Africa, United States, Australia, India, and Europe. Key focus areas of the conference were land and water management in the agricultural production value chains on one hand, and threats and opportunities associated with climate change on the other hand. Furthermore, a special emphasis was also exerted on the scaling-up of proven technologies and innovations for transformational impact on the livelihoods of African small-scale farmers.

The main focus of the CALESA final workshop was on lessons learned in the area of adapting smallholder agriculture to current climate variability and change in sub-Saharan Africa. This event aimed at providing an opportunity to scientists from the African region to share their experiences and knowledge on the one hand, and to identify gaps and priorities for future action and research on climate change adaptation on the other.

The one-day workshop consisted of one plenary and three concurrent sessions, each dealing with one of the three key components of adaptation research: planning and preparing, managing risks and opportunities, and recovering from shocks and stresses. The main sponsors of this session were ICRISAT and HAW Hamburg. The latter is the secretariat of the International Climate Change Information Programme (ICCIP).

Up-to-date information was shared among participants and mass media, including radio and television. The CALESA team members from Kenya and Zimbabwe presented ten papers including the key note address by Dr. K.P.C. Rao in the workshop. These papers covered the findings from household surveys, field experiments and the simulation analysis. The key note address by Dr. Rao dealt with direct and indirect impacts of climate variability on the productivity of smallholder agricultural systems. It also highlighted the role of climate risk which acts as a major constraint in the adoption of improved production technologies. As a result of these factors, smallholder farmers continue to rely on traditional low risk management strategies such as diversification and limited use of costly inputs such as fertilizers. The presentation also highlighted potential options which may help farmers better prepare and better manage their farms using historical and real time climate information.

The two PhD students supported by the CALESA project presented the findings from the household surveys and focus group discussions which were conducted at analogue sites in Kenya and Zimbabwe. The presentations highlighted the importance of mainstreaming gender sensitive options while developing adaptation strategies to climate variability and change. Significant differences were observed in the crop management strategies adopted, particularly between the dry analogue pair. These differences revolved around crops choices as well as soil and water management strategies. In drier areas, implications are for increased uptake of small grains. For wetter climates, soil and water management strategies are important options for smallholders. Gender issues for differently managed households seem to vary across the sites evaluated. At drier sites, gender issues include labour for production and processing of the small grains against a background of male labour migration. At wetter sites, access to draft power, labour, agricultural assets, and social and financial capital in differently managed households are important for increasing adoption of effective crop management strategies.

Trends and uncertainty in projected future climates, which were based on the MarkSim-GCM downscaled location specific climate scenarios to mid and end century periods, were presented in a paper developed by Anthony Oyoo. In addition, the main results of an ex-ante analysis which assessed climate sensitivity of management practices adopted by farmers, was discussed in a paper presented by Lucy Wangui.

Other Promotion and Dissemination Activities

Continued promotion and dissemination of the activities and outputs of the CALESA project took place not only during the lifetime of the project, but also during one year after. This was meant to facilitate and effective and adequate promotion of the final batch of transnational results and the lessons learnt of the project. The dissemination activities involved the development of a project logo, poster and brochure on one hand, and setting up and maintaining a project specific and interactive web page on the other. These elements were not only meant to communicate the project, its activities and its results, but also to catalyse concerted efforts towards linking the project with other mainstream climate adaptation activities in SSA.

To this day, at least 5 publications to scientific journals have been published. In addition, a detailed analysis of the results, at both national and transnational level, as well as adequate climate change adaptation practices and lessons learnt of the project, will be published in international renowned journals during 2014 and onwards. Furthermore, a chapter focusing on the CALESA project was published as a part of the Springer book "Experiences on climate change adaptation in Africa" on 2011, which was edited by Professor Walter Leal Filho. The CALESA project was also disseminated during the African Climate Teach-In day 2011.

Box 2 Main outputs of the CALESA project

1. Tools and approaches for delineating important crop growing areas within semi-arid tropical regions of Kenya and Zimbabwe, and their temperature analogues identified
   - Climate, soil, crop and socio-economic data necessary to characterise the target locations and establish baseline conditions collected and analysed.
   - Ex-ante assessment to quantify the current climate yield gap and risks associated with locally adopted and improved management practices at analogue locations conducted.
   - Stakeholders' perceptions about climate variability and change, and their impacts on crop management, production and other livelihood activities at analogue locations documented.
2. Through the combined use of long-term daily climate data, crop growth simulation models and participatory surveys with farmers, the implications of both current and future climate change production risks at the analogue locations identified and quantified
   - The capacity of local scientists and research organisations to build, analyse and utilize high quality climate data fostered.
   - The performance of APSIM in simulating temperature effects of growth and yield of important food crops evaluated.
   - Field studies to calibrate cultivar parameters and validate APSIM simulation of temperature effects conducted.
3. Through iterative field research both on station and in farmers' fields over more than 2 years, potential crop, soil and water management, and crop genotype adaptation options evaluated, and adaptation strategies for the target locations formulated
   - Field trials and collection of detailed data required to make a critical assessment of the performance of management options at the selected locations conducted.

- The adequacy of management options to cope with the predicted changes in climate evaluated.
- Climate change adaptation strategies for the target locations formulated.

4. Through hands-on capacity building, institutional capacity in understanding climate change impacts and developing effective adaptation responses strengthened

- 2 PhD degrees obtained.
  Research capacity-building activities in Kenya and Hamburg celebrated.
- Final CALESA workshop in Kenya conducted.
- Several articles in scientific journals and books published.
- A book focusing on the CALESA project published.

Some Lessons Learnt

Adapting Agriculture Practices to Climate Change

The studies conducted under this project provided greater insights into how farming systems under varying climatic conditions respond to the variability in the climate. Therefore, it was possible to produce good leads on the identification of options for adapting the systems to future climate changes. Some highlights from the work are as it follows:

- Even though there is no significant change in the amount of rainfall received at different locations, there are robust indications that the variability in rainfall during the main cropping season is increasing.
- Historical climate observations also revealed that temperatures at all locations are increasing. On average, temperatures increased by about 0.5 °C over the past two decades.
- While most GCMs project an increase in temperature, there are differences in the magnitude of the projected changes varying from 3 to 5 °C by end of the century. The projections in rainfall showed large variations among the GCMs. In the case of Kenya, most global projections indicate that the rainfall is going to increase during the long rainy season and decrease in the short rainy season at most locations. The differences in the predicted changes during these seasons have significant implications on how farming systems may be affected and on the options available to adapt.
- The duration of the crop generally declines with increasing temperatures. Therefore, an increase in temperature reduces the yield potential. On average, cereal yields are projected to decline by about 200 kg per ha with every one degree increase in temperature over and above the optimal temperatures.

However, there are differences in the response of different crops and varieties. Maize seems to be more sensitive to changes in temperature than sorghum. In this line, the variety WH403 is more affected than DH04.

- The data collected from various trials provided a good opportunity to calibrate and validate process-based crop simulation tools. These are meant to assess the climate sensitivity of the crops and varieties, and develop best possible options for adapting to the same. Simulation analysis with calibrated models indicated that it is possible to adapt to the future climatic conditions by making simple adjustments to the agronomic practices such as adjusting plant population, changing variety, application of fertilizers and adopting soil and water conservation measures.
- In general, communities are aware about the variability and changes in the local climatic conditions and have adapted well for the same. However, the current management is not adequate to meet the future challenges. Therefore, concerted efforts are required to promote more appropriate practices.

Farming Systems and Adoption Constraints

Adoption and adaptation are influenced by biophysical and socioeconomic environments which require appropriate policies. Some of the aspects which need to be addressed include:

- The need to address socioeconomic constraints in both male and female headed households.
- The need for mainstreaming gender in climate change adaptation as shown by high contributions of women to labour, different preferences to crops and management strategies, and in some cases, lower maize yields of maize for female headed households compared to male headed households. At drier sites there is also high male labour migration and high production of small grains which have high labour demands (production and processing) compounding the burden on women.
- The need for accompanying studies on food and nutrition, and farmers preferences, due to potential shifts in cropping landscapes. This is particularly true for drier areas.
- The need for increased adoption and adaptation through soil and water management strategies in rainfed systems. This requires technologies which are appropriate to different climates and are gender sensitive. This also requires policies which increase the availability of resources.
- Weather forecasts which provide more information with regard to the amount and distribution of rainfall may better assist farmers in planning crop management strategies.
- For livestock production, there may be need for breed improvement as well as for disease and pest control at all sites and interventions. These interventions are required because at wetter sites there is shortage of grazing land. In addition, at

drier sites the pasture is sometimes of poor quality and constrained by climate conditions.

- Exploring and optimising alternative livelihoods to supplement crop and livestock production for drier areas for 2050s climates.

Conclusions

The CALESA project was a research-for-development project which coupled integrated climate risk analyses, crop growth simulation modelling and field-based research both on-station and on-the-ground with participatory research with farmers.

The CALESA project has demonstrated that it is possible to develop robust and locally relevant adaptation strategies by combining applied research-oriented for knowledge and better technology creation with development-oriented activities for information sharing and capacity building. It has also provided a concrete contribution to evaluating the impacts of climate change and how it can be addressed by means of implementing agricultural research and development that bears in mind the needs of the poor and vulnerable.

Acknowledgments The authors wish to thank the German Agency for International Development (GIZ) for the support provided to the project CALESA.

References

Burton I, van Aalst M (2004) Look before you leap: a risk management approach for incorporating climate change adaptation into world bank operations. World Bank Monograph, Washington (DC), DEV/GEN/37 E

Chen S, Ravillion M (2007) Poverty and hunger special feature: absolute poverty measures for the developing world, 1981–2004. Proc Nat Acad Sci USA 104(43):16757

Christensen JH, Hewitson B, Busuioc A, Chen A, Gao X, Held I, Jones R, Kolli RK, Kwon W-T, Laprise R, Magaña Rueda V, Mearns L, Menéndez CG, Räisänen J, Rinke A, Sarr A, Whetton, P (2007). Regional climate projections. In: Solomon S, Qin D, Manning M, Chen Z, Marquis M, Averyt KB, Tignor M, Miller HL (eds) Climate change 2007: the physical science basis. Contribution of working group I to the fourth assessment report of the intergovernmental panel on climate change. Cambridge University Press, Cambridge

Cooper P et al (2009) Farming with current and future climate risk: advancing a hypothesis of hope' for rainfed agriculture in the semi-arid tropics. SAT eJournal 7:1–19

Cooper PJM, Dimes J, Rao KPC, Shapiro B, Shiferaw B, Twomlow S (2008) Coping better with current climatic variability in the rainfed farming systems of sub-Saharan Africa: an essential first step in adapting to future climate change? Agric Ecosyst Environ 126(1–2):24–35

DFID (2005) Climate proofing Africa: climate and Africa's development challenge. Department for International Development, London

FAO (2010) Greenhouse gas emissions from the dairy sector: a life cycle assessment

FAO (2012) Annex 2: approaches to rapid assessments of impacts of climate variability and climate change on agriculture in the project area. Incorporating climate change considerations into agricultural investment programmes: a guidance document

Galloway McLean K (2010) Advance guard: climate change impacts, adaptation, mitigation and indigenous peoples—a compendium of case studies. United Nations University—Traditional Knowledge Initiative, Darwin, Australia

ICRISAT (2008) Adaptation to climate change in the semi-arid tropics. (http://www.icrisat.org/aes-climatechange-sat.htm)

IPCC (2007) Regional climate projections In: Solomon S, Quin D, Manning M, Chen Z, Marquis M, Averyt KB, Tignor M, Miller HL (eds) Climate change 2007: the physical science basis. Contribution of working group 1 to the fourth assessment report of the intergovernmental panel on climate change, chapter 11. Cambridge University Press, Cambridge, p 996

New M, Hewitson B, Stephenson D, Tsiga A, Kruger A, Manhique A, Gomez B, Coelho C, Masisi D, Kululanga E, Mbambalala E, Adesina F, Saleh H, Kanyanga J, Adosi J, Bulane L, Fortunata L, Mdoka M, Lajoie R (2006) Evidence of trends in daily climatic extremes over southern and west Africa. J Geophys Res 111(D14102), p 11. doi:10.1029/2005JD006289

Reid H et al (2010). Community champions: adapting to climate challenges. International Institute for Environment and Development, London

Rosegrant MW, Cai X, Cline SA (2002). World water and food to 2025: dealing with scarcity. IFPRI-2020 Vision/International water management book. Washington, D.C. IFPRI 2002

Sanchez PA (2002) Soil fertility and hunger in Africa. Science 295:2019–2020

Washington R, Harrison M, Conway D, Black E, Challinor AJ, Grimes D, Jones R, Morse A, Kay G, Todd M (2006) African climate change: taking the shorter route. Bull Am Meteorol Soc 87 (10):1355–1366. doi:10.1175/BAMS-87-10-1355

Chapter 2
Improving Livelihoods in Semi-arid Regions of Africa Through Reduced Vulnerability to Climate Variability and Promotion of Climate Resilience

Stephen K. Kimani, Anthony O. Esilaba, Peterson N.M. Njeru, Joseph M. Miriti, John K. Lekasi and Saidou Koala

Abstract Climate change is expected to be one of the major threats to sustained economic growth leading to extended poverty in semi-arid regions of sub Saharan Africa (SSA). The areas of highest vulnerability are the health sector, food production, biodiversity, water resources, and rangelands. Climate change will likely create increasingly high temperatures and dry conditions across much of the globe in the next 30 years, especially along large parts of Eurasia, Africa and Australia. Many of the world's most densely populated regions will be threatened with severe drought conditions. It will likely have a profound and negative impact on livelihoods of many rural and urban communities, which could lead to changes in land use. It is estimated that the Eastern regions of Africa will experience reduced average rainfall (although some areas may experience increased average rainfall) exposing agriculture to drought stress and a rise in temperature. The situation will be worsened by the interaction of multiple stresses factors occurring at various levels, which will negatively impact agricultural productivity.

Keywords Climate change · Arid semi-arid lands · Key interventions

S.K. Kimani (✉) · A.O. Esilaba · P.N.M. Njeru · J.M. Miriti · J.K. Lekasi
Natural Resource Management Research Programme,
Kenya Agricultural Research Institute,
PO Box 57811, Nairobi, Kenya
e-mail: 2skimani@gmail.com

S. Koala
International Center for Tropical Agriculture (CIAT),
PO Box 823-00621, Nairobi, Kenya

W. Leal Filho et al. (eds.), *Adapting African Agriculture to Climate Change*,
Climate Change Management, DOI 10.1007/978-3-319-13000-2_2

Introduction

Climate change is expected to be one of the major threats to sustained economic growth that will lead to extended poverty in sub Saharan Africa (SSA). The situation is similar to other semi-arid regions of Asia. The areas of highest vulnerability are the health sector, food production, biodiversity, water resources, and rangelands. Climate change will likely create increasingly high temperatures and dry conditions across much of the globe in the next 30 years, especially along large parts of Eurasia, Africa and Australia (Cooper et al. 2013). Many of the world's most densely populated regions will be threatened with severe drought conditions. It will likely have a profound and negative impact on livelihoods of many rural and urban communities, which could lead to changes in land use. Smallholder farmers provide up to 80 % of the food in developing countries, manage the majority of the farmland, and many live in some of the most vulnerable and marginal landscapes that experience unpredictable rainfall patterns. Drylands occupy 41 % of the earth's land area and are home to 2 billion people. About 50 % of the world's livestock is supported by rangelands, and some 44 % of cultivated areas are in dry lands. However, more than 12 million hectares of arable land are lost to land degradation and desertification every year and the rate is rising as a result of climate change (van de Steeg 2012; www.unccd.int 2012). Land degradation affects 40 % of the earth's surface and damages the livelihoods of some 2 billion people living in dry lands, especially women and youth. Land degradation in dryland regions is a driver of climate change. Yet the linkages between climate change and dryland degradation have so far scarcely featured in climate change policy debate. Desertification and land degradation are reducing the capacity to sustain ecosystems and human livelihoods. Despite this, dry lands in semi-arid regions still play a major role in global agriculture production.

Agriculture directly depends on climatic factors for crop and livestock production. Agricultural practices are also indirectly affected by landscape and environmental changes brought about by climate change. It is the SSA countries whose economies heavily depend on agriculture (cultivation of crops and livestock production) and forestry that are particularly vulnerable to climate change and variability, and will bear about 80 % of the effect (Mandelson 2006). Several industries and investments in SSA are agro-based. Declining agricultural output is likely to affect value chains. Though of smaller magnitude, a reduction in agricultural GDP can affect the rate of industrialization and the overall development process of many SSA countries and constrain creation of non-farm rural and urban employment opportunities through backward and forward linkages to service and manufacturing sector activities (Hanmer and Naschold 2000; Kanwar 2000; Kogel and Furnkranz-Prskawetz 2000).

It is projected that several ecosystems will experience a number of climate related stresses. It is estimated that especially the Eastern regions of Africa will experience reduced average rainfall (although some areas may experience increased average rainfall) exposing agriculture to drought stress and a rise in temperature

(Cooper et al. 2013). The situation will be worsened by the interaction of multiple stresses factors occurring at various levels. For example heat and drought stresses often occur simultaneously. Combined, these affects will negatively impact agricultural productivity. According to the Secretary General of the UN, Ban Ki-Moon, Continued land degradation—whether from climate change, unsustainable agriculture or poor management of water resources, is a threat to food security, leading to starvation among the most acutely affected communities and robbing the world of productive land (Pender et al. 2009).

In addition, many policy makers in governments are unaware of this long-term climatic impact, often leading to land use changes. Governments have low or no adaptive strategies or capacity to make people aware of climate change and climatic impacts in the long-term. Climate change is expected to reduce yields of major crop staples and will condemn portions of currently cultivated land into unsuitable status for cultivation across many parts of SSA. It is estimated that yields of tropical grain crops are expected to be reduced by 5–11 % by the year 2020 and by 11–46 % by 2050 (Rosenzweig and Parry 1994; Schlenker and Lobell 2010; Blanc 2012), negatively impacting on the small scale farmers who solely rely on rain-fed agriculture for their livelihood. Projected GDP losses in SSA are estimated to range between 0.2 and 2 % by 2100 (Tol 2002). National adaptation and mitigation planning is urgently needed. In addition, low agricultural productivity has increased pressure on traditional grazing lands by expanding cultivation into rangelands. This has lead to more rapid degradation of rangeland ecosystems.

If the ultimate effect of climate change and variability is not attended to, it may contribute to political instability and migration, at both intra-and regional levels. A recent survey conducted by IFPRI with the support of World Bank identified migration as one of the major adaptation strategies among the communities in semi-arid environments (World Bank 2000; IFPRI 2010). A number of regions/sub-regions across SSA have just emerged from, or are experiencing conflicts. A new wave of ecological refugees will spark a series of conflicts among communities and complicating the development agenda of several SSA countries, if there is no action to reduce the effects of climate change now (UNFCCC 2007). A good baseline study to complement earlier efforts on the possible effects of climate change in vulnerable, poor countries is therefore urgently needed, before sustainable mitigation measures can be implemented that will stabilize or stimulate economic growth in the long-term.

Adaptation and mitigation strategies are two general responses to manage effects of climate change and variability. Although adaptation represents the best coping option against agricultural output reduction and hence resulting in improved livelihood of small holder farmers; mitigation actions will contribute to global efforts of greenhouse gas emissions reduction, sequestration of carbon as practical measures for climate change recovery, taking advantage of the carbon storage capacity of tropical environment and improving ecosystem services of the natural resource (FAO 2001; World Bank 2012). The African Development Bank (AfDB) for example has developed their Climate Risk Management and Adaptation (CRMA) strategy which outlines key priority areas of intervention in order to manage the risks posed by climate change.

The goal as stated in the strategy document is "to ensure progress towards eradication of poverty and contribute to sustainable improvement in people's livelihoods taking into account CRMA". Specific objectives of the CRMA are to reduce vulnerability within the Regional Member Countries (RMCs) to climate variability and promote climate resilience in past and future Bank financed development investments making them more effective. This will then be used to build capacity and knowledge within the RMCs to address the challenges of climate change and ensure sustainability through policy and regulatory reforms."

To achieve these objectives the AfDB considered supporting three areas of intervention namely:

I. "*Climate Proofing*" investments to ensure that development efforts are protected from negative impacts of climate change, climate variability and extreme weather events.
II. Support the development of *Policy, Legal and Regulatory Reforms* which creates an enabling environment for the implementation of climate risk management and adaptation interventions.
III. *Knowledge Generation and Capacity Building* for local farmers, investors, extension agents, district executives or policy makers to help mainstream climate change and manage climate risks.

Over time, some SSA countries have also developed their climate change policy plans including National Adaptation Programme Actions and National Appropriate Mitigation Actions. Investments are needed in building up assets, implement recommended promising technologies/practices (e.g. water harvesting, storage, irrigation system, introduction of drought tolerant high yielding crops, value addition) and improving risk management capacity. As acknowledged by Stern (2006), the biggest threat climate change poses to economic growth is the use of inefficient mitigation and adaptation policies and practices (Stern 2006). To improve the efficiency of these actions, it is important that they are based on accurate spatio-temporal impact diagnosis, and supported by a greater public understanding of these strategies and individual roles.

Unfortunately significant gaps of knowledge exist on the most appropriate interventions to use. Many actors (government, agencies and investors) are asking what options exists and which should be implemented to improve the livelihoods of the rural poor. The bottom-line costs and/or benefits of these interventions need to be known if they are to be planned and implemented and investments sources to support their development. Therefore, there is need to continue grappling with ideas, as well as ways and means which can contribute towards improved incomes and livelihoods in semi-arid regions of Africa through reduced vulnerability to climate variability and promotion of climate resilience in development investments, enhancing biodiversity, increasing yields and lowering greenhouse gas emissions.

Key Interventions

I. There is need first of all to quantify vulnerability to climate change, adaptation approaches by systematic monitoring across landscapes and identify barriers to successful mainstreaming of these adaptation measures in country national plans.
II. This needs to be followed by developing, promoting and adapting site specific mitigation/adaptation measures for various crop-livestock land use systems
III. Thirdly, development of Policy, Legal and Regulatory frameworks in order to create an enabling environment for the implementation, promotion and scaling of climate risk management and adaptation interventions needs to be emphasized.
IV. Sub-Saharan Africa also needs to build Capacity to mainstream climate change and manage climate risks for various land use systems.

Expected Outputs from the Interventions

I. Vulnerability to climate change, and climate change impacts quantified and mainstreamed in country national development, management and policy plans.
II. Site specific adaptation and mitigation measures for various crop-livestock land use systems developed and promoted in arid and semi-arid areas
III. Policy, Legal and Regulatory frameworks developed and promoted and scaled in order to create an enabling environment for the implementation of climate risk management and adaptation interventions.
IV. Capacity to mainstream climate change and management of climate risks for various land use systems by national scientists, agriculturalist, environmental experts and policy makers developed.

Expected Outcomes of the Interventions

I. Implemented country development plans embrace, adopt and mainstream climate change impacts at national and regional/county levels
II. Improved incomes and livelihoods in semi-arid regions through yield increases in crop-livestock production systems, reduced crop and livestock losses and reduced greenhouse gas emissions as a result of implemented adaptation and mitigation measures for climate change.
III. Action plans on climate risk management and adaptation interventions implemented in project countries.
IV. Trained national scientists, agriculturalists, environmental experts and policy makers mainstream and implement climate change and management of climate risks for various land use systems in project countries.

Expected Impact

I. Impact on livelihood: Improved household incomes and livelihoods, improved national GDPs. The ultimate beneficiaries are resource poor farmers and other members of the rural and peri-urban poor associated with the agricultural sector. These benefits will be realized through reduced vulnerability, raised adaptive capacity and higher income.
II. Impact on food security benefits on rural and urban populations, and
III. Impact on environmental health and carbon storage at both local on global public goods.
IV. Although the notion of securing win-win-win outcomes for these dimensions is appealing (Global Donor Platform 2009; FAO 2009), we have to recognize the possibility of trade-offs among these dimensions (Campbell et al. 2009; FAO 2011).

Proposed Theory of Change

Reduced vulnerability to climate variability and change and promotion of climate resilience requires development of investments in support of reducing poverty, enhancing biodiversity, increasing yields and lowering greenhouse gas emissions. This will be achieved through the following preconditions. Firstly, the projects should undertake a quantification of vulnerability to climate change, adaptation approaches and identify barriers to success mainstreaming of these adaptation measures in country national plans. This will include identifying key metrics of vulnerability in crop-livestock systems, rangelands, and agricultural systems. Secondly, the formulated projects will need to develop, promote and scale climate change adaptation and mitigation measures for various crop-livestock and land use systems in the arid and semi-arid areas. Thirdly the formulated projects will require to undertake activities that promote development of policy, legal and regulatory frameworks in order to create an enabling environment for the implementation of climate risk management and adaptation interventions. Finally the proposed projects will need to build capacity to mainstream climate change, manage climate risks for various land use systems as well as mainstream gender along selected agricultural product value chains.

Indicators for these preconditions which will be used to assess the performance of the interventions will be an inventory of vulnerable groups, adaptation approaches, and barriers to mainstreaming these approaches in sub-Saharan Africa. A wide range of climate change adaptation and mitigation measures will need to be promoted and scaled. These will include improved land use planning, improved agricultural practices that enhance soil carbon stocks, better livestock management, use of drought tolerant crops, improved irrigation and water use efficiency, and rain water harvesting. A set of policy documents will then need to be developed for the participating countries from which action plans will be generated and implemented. Researchers, extension workers, policy makers and other relevant stakeholders will be trained and then subsequently use this knowledge and share it with other parties to achieve project overall goal. Training will include simulation modeling, greenhouse gas emission measurements, carbon stocks measurements, and participation in carbon credits market for participating countries.

Expected Impact pathway: This is explained in the diagram below.

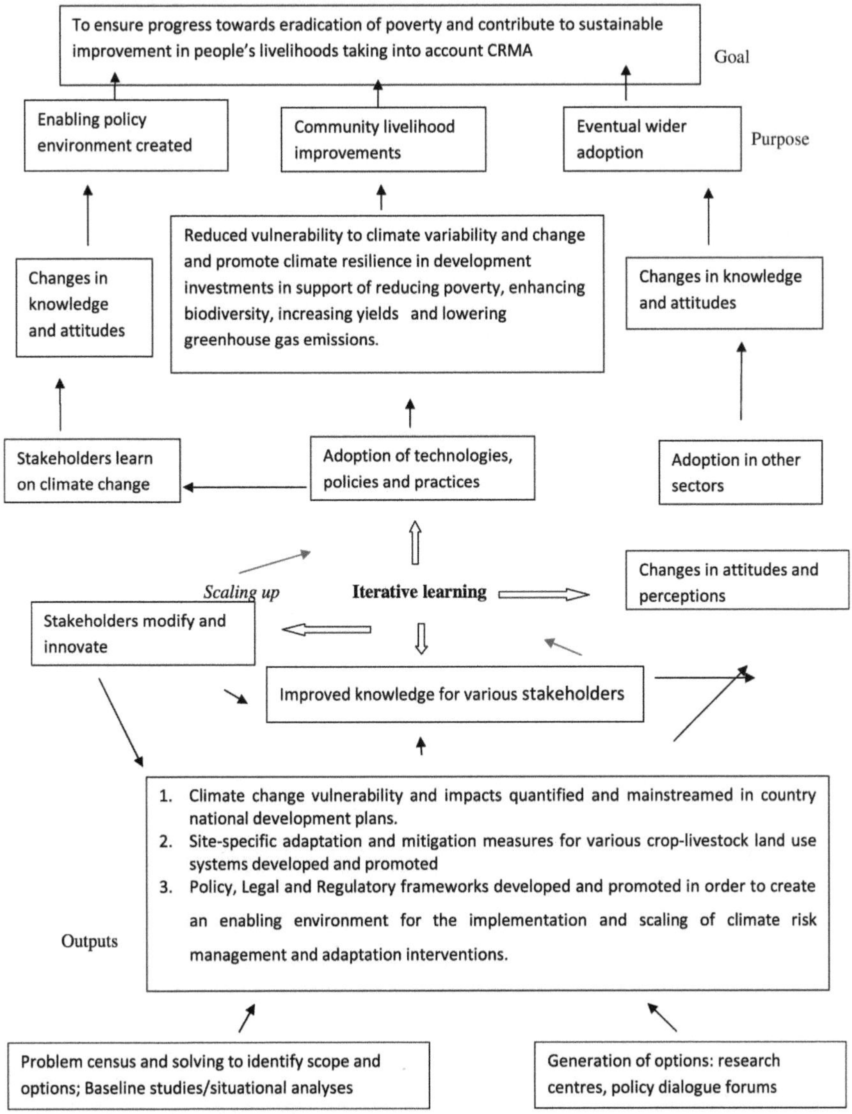

Partnerships

Building strong partnerships will form an essential component of implementing projects using this approach. Essentially, this may be worked out as consortia with complementary partnerships in order to ensure the long-term impact of this initiative and provide the greatest opportunity for knowledge transfer. Partners will include, but are not limited to:

I. Consultative Group of International Agricultural Research (CGIAR) Centres
II. Non-governmental and community based organizations for rural development
III. National agricultural research systems
IV. Ministries responsible for National Adaptation Programmes of Action (NAPAs) and Nationally Appropriate Mitigation Actions (NAMAs)
V. Regional clean development brokers
VI. Climate change Units
VII. Gender mainstreaming experts
VIII. Private sector
IX. Development partners

References

Blanc E (2012) the impact of climate change on crop yields in Sub-Saharan Africa. Am J Clim Change 1:1–13. doi:10.4236/ajcc.2012.11001 [Published Online March 2012 (http://www.SciRP.org/journal/ajcc)]

Campbell A, Kapos V, Scharlemann JPW, Bubb P, Chenery A, Coad L, Dickson B, Doswald N, Khan MSI, Kershaw F, Rashid M (2009) Review of the literature on the links between biodiversity and climate change: impacts, adaptation and mitigation. Secretariat of the Convention on Biological Diversity, Technical series no. 42, Montreal, 124 p

Cooper PJM, Stern RD, Noguer M, Gathenya JM (2013) Climate change adaptation strategies in Sub-Saharan Africa: foundations for the future. In: Singh BR (ed) Climate change—realities, impacts over ice cap, sea level and risks

FAO (2001) Soil carbon sequestration for improved land management. World Soil Resources Report No. 96, Rome

FAO (2009) Profile for climate change food and agriculture organization of the United Nations. Food and Agriculture Organization of the United Nations, Rome

FAO (2011) Adapt framework programme on climate change adaptation. Food and Agriculture Organization of the United Nations, Rome

Global Donor Platform for Rural Development (2009) Guidelines for donor support to CAADP process at a Country-Level. c/o Federal Ministry for Economic Cooperation and Development. (BMZ) Dahlmannstraße 4, 53113 Bonn, Germany

Hanmer L, Naschold F (2000) Attaining the international development targets: will growth be enough? Dev Policy Rev 18(1):11–36

IFPRI (2010) Strategies for adapting to climate change in rural Sub-Saharan Africa. IFPRI discussion paper 01013 July 2010

Kanwar S (2000) Does the dog wage the tail or the tail the dog? cointegration of indian agriculture with nonagriculture. J Policy Model 22(5):533–556

Kogel T, Furnkranz-Prskawetz A (2000) Agricultural productivity growth and escape from the malthusian trap. CEPR working paper no. 2485, June
Mandelson P (2006) Report on trade and climate change. EU, Brussels
Pender J, Ringler C, Magalhaes M (2009) The role of sustainable land management (SLM) for climate change adaptation and mitigation in Sub-Saharan Africa (SSA). Regional Sustainable Management Publication, Terr Africa, 110 p
Rosenzweig C, Parry ML (1994) Potential impacts of climate change on world food supply. Nat 367(13):133–138
Schlenker W, Lobell DB (2010) Robust negative im-pacts of climate change on african agriculture. Environ Res Lett 5(1):1–8. doi:10.1088/1748-9326/5/1/014010
Stern R (2006) Review on the economics of climate change. Grantham Research Institute on Climate Change and Environment, UK, 700 p
TOL RJS (2002) Estimates of the damage costs of climate change. Part II. Dynamic Estimates. Environ Resour Econ **21:** 135–160, 2002 (© 2002 Kluwer Academic Publishers. Printed in the Netherlands, 135)
UNFCCC (2007) Adaption under the frameworks of the CBD, the UNCCD and the UNFCCC. Joint Liaison Group of the Rio Conventions. Available at http://unfccc.int/resource/docs/publications/adaptation_eng.pdf
van de Steeg J (2012) Livestock and climate change in the near east region. Measures to adapt to and mitigate climate change. Food and Agriculture Organization of the United Nations. Regional office for the near east, Cairo
World Bank (2000) Can Africa claim the 21st century? World Bank, Washington, DC
World Bank (2012) Carbon sequestration in agricultural soils. The World Bank, Washington DC

Chapter 3
Climate Change Adaptation Planning in Kenya: Do Scientific Evidences Really Count?

Kizito Kwena, William Ndegwa, Anthony O. Esilaba, Sospeter O. Nyamwaro, Dickson K. Wamae, Stella J. Matere, Joan W. Kuyiah, Reuben J. Ruttoh and Anthony M. Kibue

Abstract The aim of this study is to assess the extent to which scientific information has been used to inform climate change adaptation policies, plans and strategies in Kenya; and also to assess the effectiveness of existing platforms for sharing climate change information in the country. Two major policy documents guiding climate change adaptation planning in Kenya, the National Climate Change Response Strategy (NCCRS) and the National Climate Change Action Plan (NCCAP), were analysed for use of scientific information in their formulation through literature review; and interviewing policy makers using an open-ended questionnaire to determine the extent to which they accessed and applied scientific-based evidence of climate change impacts in development planning. Both documents, the NCCRS and NCCAP, made fairly good use of evidence contained in technical reports, especially the UNFCC, World Bank and FAO reports. However, they made very minimal, less than 20 %, reference to the hard scientific facts offered by journals, books and workshop proceedings. Similarly, only about 6 % of the respondents used the climate change information to develop mitigation and adaptation plans, training curricula, and Research and Development programs. The rest, over 76 %, rarely

---

K. Kwena (✉) · R.J. Ruttoh
KARI Katumani, P.O. Box 340-90100, Machakos, Kenya
e-mail: kwenakizito@yahoo.com

W. Ndegwa · A.M. Kibue
Kenya Meteorological Department, P.O. Box 30259-00100, Nairobi, Kenya

A.O. Esilaba
KARI Headquarters, P.O Box 57811-00200, Nairobi, Kenya

S.O. Nyamwaro · D.K. Wamae
KARI Muguga North, P.O. Box 032-00902, Kikuyu, Kenya

S.J. Matere
KARI Muguga South, P.O. Box 30148-00100, Nairobi, Kenya

J.W. Kuyiah
Ministry of Agriculture, P.O. Box 30028-00100, Nairobi, Kenya

W. Leal Filho et al. (eds.), *Adapting African Agriculture to Climate Change*,
Climate Change Management, DOI 10.1007/978-3-319-13000-2_3

used it for planning purposes. This could be attributed to limited knowledge of appropriate methodology to distil relevant decision-relevant information from the spectrum of available information on climate change projections, availability of the information in user-unfriendly formats, and lack of information sharing protocols. There is need to reverse this trend. Most respondents (42 %) preferred the agricultural extension system in delivering climate change information. This was followed by stakeholders meetings with 29 % of the respondents' preference, conferences and workshops with 5 %, media (4 %), and climate change networks and internet with less than 1 % each. However, the national agricultural system is severely constrained by staff and facilities, and is therefore very limited in its reach. There is therefore need to strengthen it and also take full advantage of recent advances in ICT if the war against climate change is to be won. Meanwhile, majority of the respondents (50 %) were ignorant of the existence of any climate change databases. But about 17 % of the respondents were aware of and accessed databases hosted by Consultative Group on International Agricultural Research (CGIAR) and other international research centres. Another 10 % of the respondents relied on databases managed by donor agencies whilst about 8 % of the respondents each accessed databases established by Government Departments and National Agricultural Research Institutions (NARIs). Finally, about 7 % of the respondents relied solely on the FAO-based databases. The preference by respondents for databases managed by CGIAR centres may be attributed to the richness and accessibility of these databases due to very active participation of these centres in climate change research. There is need to enrich NARIs databases and those of Government Departments and make them more accessible to enhance sharing and application of climate change information by policy makers and other stakeholders.

Keywords Climate change adaptation · Scientific information · Policy documents · Policy makers

Introduction

Agriculture is the mainstay of Kenya's economy. It accounts for over 26 % of Kenya's GDP, 60 % of her export earnings and employs over 80 % of her workforce (GoK 2004, 2007). The crops sub-sector contributes 60 % of the agricultural GDP, while livestock and fisheries sub-sectors contribute the remaining 40 % (GOK 2010). However, like the rest of Sub-Saharan Africa, Kenya's agriculture is mainly rain-fed and therefore highly vulnerable to climate change. Already, per capita food production in the country has declined over the past two decades, contrary to the global trend. For instance, maize yields have fallen from 2 to 0.5 t/ha over the past 10 years resulting in widespread malnutrition, a recurrent need for emergency food supply and an increasing dependence on food imports (Hassan et al. 1998). Estimates available indicate that about 50.6 % of the Kenyan population lacks access to adequate food and lives in abject poverty, and this figure is bound to increase given the current population growth rate of 3 % (GoK 2004). Already, the Government has spent over ksh. 20 billion in the past five years to feed 3.5–4.5 million people annually (GoK 2009, 2010).

The low productivity has been attributed partly to declining soil fertility, but mainly to climate variability, especially in arid and semi-arid areas (ASALs), which account for over 80 % of Kenya's total area. Rainfall in these areas is low (300–500 mm annually), highly variable and unreliable for rain-fed agriculture and livestock production (Herrero et al. 2010; WRI 2007). The situation is bound to worsen with the expected change in climate.

Climate change projections indicate that Kenya's temperatures and rainfall variability will increase by about 4 °C and 20 %, respectively, by 2030. Thus, droughts and floods are bound to be more frequent and severe in both the ASALs and high potential areas. These are highly likely to exacerbate the already precarious food, water and energy situation in the country; and cause severe shortage of other essential basic commodities and long term food insecurity if left unmanaged. Vulnerability mapping studies conducted in the East African region predict that yields of major staples in ASALs and coastal areas will decrease by 20–50 % (Thornton et al. 2009).

Thus, climate change is likely to expose more people more frequently and for longer periods to threats to their livelihoods arising from extreme weather events. Consequently, more households across the country will be trapped in chronic food insecurity and chronic poverty. The Government has formulated the National Climate Change Response Strategy (NCCSR) and National Climate Change Action Plan (NCCAP) to guide adaptation planning in the country to minimize the negative impacts and optimize on the opportunities presented by climate change. However, for the proposed measures to be effective and widely adopted they have to be supported by credible estimates of their cost, impact, and economic benefits, something that has been lacking in most National Adaptation Plans of Action (NAPAS) and other climate change adaptation policy documents in the region. This study therefore sought to assess the extent which scientific information has been used to inform climate change adaptation policies, plans and strategies in Kenya. It also sought to assess the effectiveness of existing platforms for sharing climate change information in the country.

Materials and Methods

The study was conducted in Kenya and had two components. The first part involved interrogating two major policy documents guiding climate change adaptation planning in the country, namely, the National Climate Change Response Strategy (NCCRS) and the National Climate Change Action Plan (NCCAP) for use of scientific information in their formulation.

The second component involved interviewing policy makers to determine the extent to which they applied science-based evidences of climate change impacts in development planning. The survey was conducted among Government Departments involved in climate change work, National Agricultural Research Institutions (NARIs), public universities, Non-Governmental Organizations (NGOs) and policy makers. Among the public universities covered by the study were the University of Nairobi, and Moi, Maseno, Egerton and Kenyatta Universities, Jomo Kenyatta University of Agriculture and Technology (JKUAT) and Masinde Muliro University of Science and Technology.

About 90 respondents drawn from these institutions were interviewed using an open-ended interviewing questionnaire. The questionnaire was designed to capture information on existing communication channels and application of estimates of climate change impacts in the development planning process. The data were coded, entered, cleaned and analyzed using the SPSS computer program. The results are presented both graphically and by descriptive statistics.

Results and Discussions

Platforms for Knowledge Sharing on Climate Change

One of the weaknesses of climate change research in Kenya and the region as a whole has been lack of proper communication between researchers and other stakeholders. Consequently, most of the research findings rarely get to policy makers and other end-users, and researchers hardly get any feed-back from them. This study sought to identify existing knowledge sharing platforms and how researchers are making use of them to convey climate change information. Most respondents (42 %) preferred the agricultural extension system in delivering climate change information. This was followed by stakeholders meetings with 29 % of the respondents' preference, conferences and workshops with 5 %, media (4 %), and climate change networks and internet with less than 1 % each (Fig. 3.1). However, the national agricultural system is severely constrained by staff and facilities, and is therefore very limited in its reach. There is therefore need to strengthen it and also take full advantage of recent advances in ICT if the war against climate change is to be won.

On whether climate change information was being used or not, this study established that only about 6 % of the respondents used it to develop mitigation and adaptation plans, training curricula, and R&D programs. The rest majority of the respondents (over 76 %) did not use the information for anything else including planning (Fig. 3.2). The low level use of climate change information may be attributed to limited knowledge of appropriate methodologies to distil relevant decision-making information from the spectrum of available information on climate

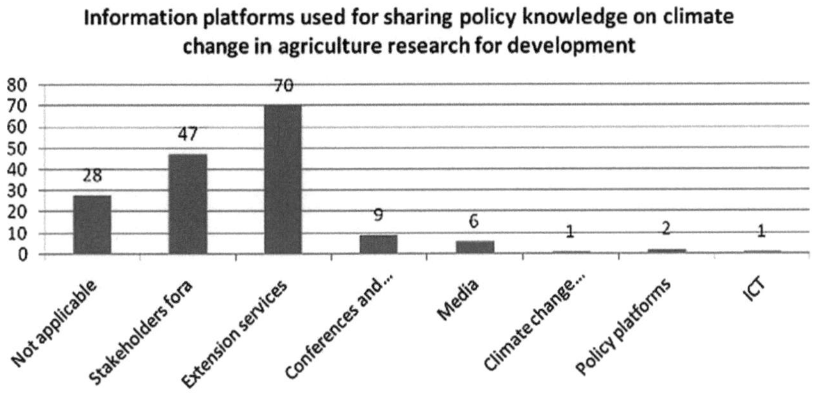

Fig. 3.1 Knowledge sharing platforms

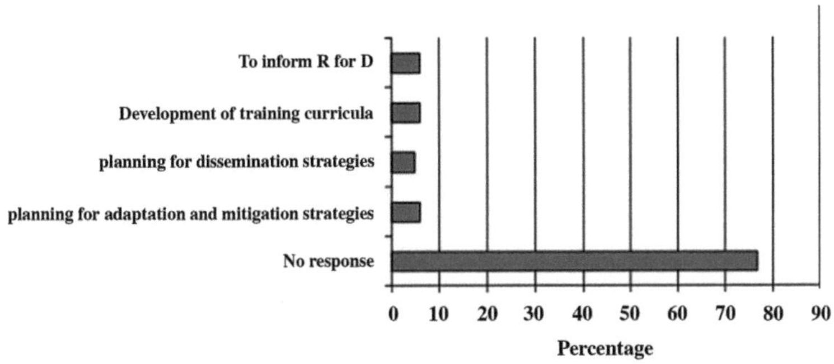

Fig. 3.2 Utilization of climate change information

change projections, availability of the information in user-unfriendly formats, and lack of information sharing protocols. There is need to reverse this trend.

The study also sought to know the climate change databases that researchers and other stakeholders were aware of and therefore accessing to share knowledge and experiences on climate change. Majority of the respondents (50 %) were ignorant of the existence of any climate change databases. But about 17 % of the respondents were aware of and accessed databases hosted by Consultative Group on International Agricultural Research (CGIAR) and other international research centres such as International Livestock Research Institute (ILRI), International Crop Research Institute for Semi-Arid Tropics (ICRISAT) and International Centre for Research in Agroforestry (ICRAF). Another 10 % of the respondents relied on databases managed by donor agencies such as International Development and Research Centre of Canada (IDRC), Department for International Development (DfID) of the United Kingdom, and the Association for Strengthening Agricultural Research in Eastern and Central Africa (ASARECA), whilst about 8 % of the respondents each accessed

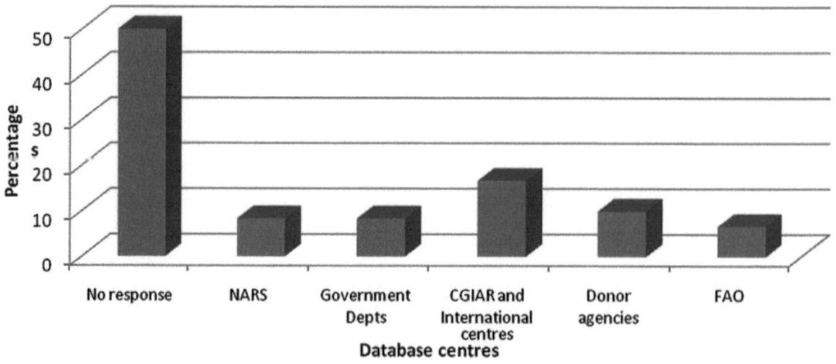

Fig. 3.3 Climate change databases

databases established by Government Departments and National Agricultural Research Institutions (NARIs). Finally, about 7 % of the respondents relied solely on the FAO-based databases. The preference by respondents for databases managed by CGIAR centres may be attributed to the richness and accessibility of these databases due to very active participation of these centres in climate change research. There is need to enrich NARIs databases and those of Government Departments and make them more accessible to enhance sharing and application of climate change information by policy makers and other stakeholders (Fig. 3.3).

Application of Scientific-Based Evidence in Adaptation Planning Process

One of the objectives of this study was to examine the extent to which scientific-based evidence of climate change impacts have been applied in adaptation planning. To do this, the study scrutinized two key policy documents that guide adaptation planning in the country, namely the National Climate Change Response Strategy (NCCRS) and National Climate Change Action Plan (NCCAP), for any use of scientific information in their formulation. Both documents made fairly good use of evidence contained in technical reports, especially the UNFCC, World Bank and FAO reports. However, they made very minimal, less than 20 %, reference to the hard scientific facts offered by journals, books and workshop proceedings (Table 3.1).

Table 3.1 Utilization of scientific information in formulation of climate change policies and strategies

Category	NCCRS		NCCAP	
	n	%	n	%
Synthesis/technical reports	27	34.6	36	53.7
Concept/working/discussion papers	12	15.4	0	0
Refereed journals	9	11.5	6	9
Books	5	6.4	9	13.4
Workshop proceedings and presentations	12	15.4	2	3
Government policies and other documents	7	9	12	17.9
Newspaper and newsletter articles	6	7.7	2	3
Total	78	100	67	100

Conclusion and Recommendation

Generally, application of scientific information in adaptation planning is very low probably due to lack of methodology to distil decision-relevant information from the spectrum of available information on climate change projections, availability of the information in user-unfriendly formats, and lack of information sharing protocols. There is need to develop effective communication channels to facilitate information sharing between researchers and other stakeholders, especially farmers and policy makers.

Acknowledgment The authors are grateful to IDRC through the project "Enhancing Climate Change Adaptation in Agriculture and Water Resources in the Greater Horn of Africa (ECAW)" for funding this study. We are also grateful to our numerous respondents for their cooperation and willingness to participate in this study.

References

Government of Kenya (GoK) (2004) Strategy for revitalizing agriculture, 2004–2014. Ministry of Agriculture and Ministry of Livestock and Fisheries Development, Nairobi, Kenya

Government of Kenya (GoK) (2007) Kenya vision 2030. Ministry of Planning and National Development, Kenya

Government of Kenya (GoK) (2009) Economic review of agriculture. Ministry of Agriculture, Kenya

Government of Kenya (GoK) (2010) The agriculture sector development strategy (ASDS). Ministry of Agriculture, Kenya

Hassan RM, Murithi FM, Kamau G (1998) Determinants of fertilizer use and the gap between farmers' maize yields and potential yields in Kenya. In: Hassan RM (ed) Maize technology development and transfer. CAB International, pp 137–161

Herrero M, Ringler C, van de Steeg J, Thornton P, Zhu T, Bryan E, Omolo A, Koo J, Notenbaert A (2010) Climate variability and climate change and their impacts on Kenya's agricultural sector. ILRI, Kenya

Thornton PK, Jones PG, Alagarswamy G, Andresen J (2009) Spatial variation of crop yield response to climate change in East Africa. Glob Environ Change 19:54–65

WRI (World Resources Institute) (2007) Nature's benefits in Kenya, an atlas of ecosystems and human well-being. WRI Department Resource Surveys and Remote Sensing, Washington, DC, USA; Ministry of Environment and Natural Resources, Nairobi, Kenya; Central Bureau of Statistics, Ministry of Planning and National Development Kenya; and ILRI, Kenya. Available at http://www.wri.org/

Chapter 4
Situation Analysis of Climate Change Aspects in Kenya

S.O. Nyamwaro, D.K. Wamae, K. Kwena, A.O. Esilaba, W. Ndegwa, S.J. Matere, K.J. Wasswa, R. Ruttoh and A.M. Kibue

Abstract Given that climate change and variability have become one of the greatest threats to food security and livelihoods, a baseline study and some literature synthesis were conducted to understand the current situation of CC scenarios in Kenya. The study sought to determine the current status of CC projects that have been undertaken in Kenya in the past five years. Major CC themes and sensitive productive sectors to CC were conceptualized in which the study was based. The baseline survey targeted key informants in academic, research and policy arenas. It was observed that adaptation, mitigation and capacity building accounted for 60, 17 and 23 % of the projects sampled. Agricultural sector (crops) accounted for most of CC projects, accounting for 36 % as well as 40 % of all projects on adaptation. Agriculture, livestock and environment sectors accounted for 30 % each of the mitigation projects. It is established that most projects undertaken in Kenya on CC arena have been on adaptation, capacity building and mitigation. CC projects undertaken in Kenya were in agriculture and livestock sectors. Although considerable efforts appear to have been put in adaptation to CC, more needs to be done, especially in agriculture and water sectors, which are important in Kenya's economy.

S.O. Nyamwaro (✉) · D.K. Wamae
KARI Muguga North, P.O. Box 032-00902, Kikuyu, Kenya
e-mail: snyamwaro344@gmail.com

K. Kwena · R. Ruttoh
KARI Katumani, P.O. Box 340-90100, Machakos, Kenya

A.O. Esilaba
KARI Headquarters, P.O Box 57811-00200, Nairobi, Kenya

W. Ndegwa · A.M. Kibue
Kenya Meteorological Department, P.O. Box 30259-00100, Nairobi, Kenya

S.J. Matere
KARI Muguga South, P.O Box 57811-00200, Nairobi, Kenya

K.J. Wasswa
Ministry of Agriculture, P.O. Box 30028-00100, Nairobi, Kenya

W. Leal Filho et al. (eds.), *Adapting African Agriculture to Climate Change*, Climate Change Management, DOI 10.1007/978-3-319-13000-2_4

Keywords Climate change adaptations · Mitigation · Capacity building · Situation analysis

Introduction

Climate change (CC) is a serious threat to agricultural productivity in regions that are already food insecure. Evidence of crop yield impact in Africa and South Asia resulting from CC is clearly witnessed in wheat, maize, sorghum and millet, and is unclear, absent or contradictory in rice, cassava and sugarcane (Knox et al. 2012). It is projected that by 2050 the world will have to increase agricultural production to feed a projected nine billion people against changing consumption patterns, impacts of CC and growing scarcity of water and land (Beddington 2010). Sub-Saharan Africa (SSA) is reported as the most vulnerable region to CC and variability (Slingo et al. 2005). This is partly because SSA maintains the highest proportion of malnourished populations with substantial portion of its national economies dependent on agriculture (Schlenker and Lobell 2010; Kpadonou et al. 2012) and most of its available water resources (85 %) used for agriculture (Downing et al. 1997). Farming techniques in SSA have also not kept abreast with modern technology, with a majority of its land arid and semi-arid, and smallholder systems that have limited capacity to adapt dominating agricultural landscape (Müller et al. 2011). Hence development externalities associated with CC will be most felt in Africa. Some CC extremes such as seasonal droughts and floods are already undermining economies and prosperity of the SSA and its people.

In Kenya, the effects of climate change and variability (CCV) are becoming more conspicuous and real given that their impacts are already affecting ecosystems, biodiversity and people. Climate change extremes such as unpredictably more frequently occurring droughts and flooding are already undermining the economies and prosperity of Kenya and the Greater Horn of Africa. Agriculture and water resources are among key sectors that are getting affected most by the impacts of CC scenarios. Climate change has the potential to slow down economic development of Kenya and many other countries.

Currently there is growing evidence of increased climate change and variability (CCV) in Kenya, leading to more than one drought every five years. This is causing substantial and irreversible decreases in productive sectors, particularly in livestock numbers in the arid and semi-arid lands (ASALs) of Kenya (MacMillan 2011). The droughts and floods expose the livestock industry to serious vulnerability and myriads of problems including livestock deaths, high malnutrition rates and diseases incidences. During the 2009 drought, Kenyan pastoralists lost more than 50 % of their herds; 81 and 64 % of their cattle, and sheep and goats respectively (African Conservation Centre 2012; Mutimba et al. 2010).

Global circulation models predict that by year 2100, climate change (CC) will increase temperatures by 4 °C leading to serious crop failures, reduced water and forage availability, and increased livestock mortalities and loss of livelihoods (Nanyingi et al. 2012). Similarly, Knox et al. (2012) projected impacts of climate change on the yield of eight major crops in Africa and South Asia showing that projected mean change in yield of all crops is −8 % by the 2050s in both regions. Across Africa, mean yield changes of −17 % (for wheat), −5 % (for maize), −15 % (sorghum) and −10 % (millet) were estimated. It is also predicted that potential cost to Africa due to CC dynamics will reach about US$10 billion per year by 2030 (PACJA 2009). Hence, mainstreaming adaptation capacity in Kenya and African development policy, planning and investment processes is absolutely relevant. In spite of uncertainties surrounding CC projections, adaptation planning remains a relevant integral component of development and investments.

In order to provide practical roadmaps for future adaptation investments, programs for adaptation actions such as the National Adaptation Programs of Action need strengthening. One way of doing this is through conducting economic analyses of adaptation investments that are informed by credible and impartial scientific assessments of climate change (CC) impacts. Towards tackling economic analyses of adaptation options in Kenya, it became necessary to understand the current situation analysis of CC scenarios within the country. Major CC themes and sensitive productive sectors to CC were thus conceptualized in which the analysis was based.

Material and Methods

A baseline survey was undertaken to determine the current status of CC projects that have been undertaken in Kenya during the past five years. Ninety respondents drawn from universities, government departments, national research institutions and non-governmental organizations (NGOs) were interviewed using a structured open-ended questionnaire. The survey targeted key informants in academic, research and policy arenas. Most respondents however came from academic institutions (universities) and a few researchers and policy planners. The collected data were coded, entered, cleaned and analyzed using the SPSS Version 18 software.

Results and Discussion of the Projects Survey

Projects in Selected CC Thematic Areas

Given that CC effects and impacts are being prevented, stopped and tolerated from happening and/or proceeding further, three major thematic areas/scenarios were conceptualised on what actions are being taken against CC in Kenya. The commonest actions being undertaken in Kenya were adaptation, mitigation and capacity building, which were regarded as the major CC thematic areas.

Adaptation to CC (or global warming) involves acting to tolerate effects of global warming, an adjustment in natural or human systems in response to actual or expected climatic stimuli or their effects. Adaptation measures may include prevention, tolerance or sharing of losses, changes in land use or activities, changes of location, and restoration. In contrast, climate change (CC) mitigation is action to decrease the intensity of radiative forcing in order to reduce the effects of global warming (Marland et al. 2007; IPCC 2007; GoK 2010). Climate change mitigation scenarios involve reductions in the concentrations of greenhouse gases, either by reducing their sources or by increasing their sinks (Molina et al. 2009). The UN defines mitigation as a human intervention to reduce the sources or enhance the sinks of greenhouse gases. Mitigation include using fossil fuels more efficiently for industrial processes or electricity generation, switching to renewable energy (solar or wind power), improving insulation of buildings, and expanding forests and other 'sinks' to remove greater amounts of CO_2 from the atmosphere (UNFCCC 1997). It is important to note that adaptation and capacity building are more implementable at the micro level, while mitigation at the macro level.

Effective responses to CC combine both adaptation and mitigation strategies. There are clear complementarities in applying both mitigation and adaptation aspects to CC, although they differ in important respects. Benefits from mitigation are expected to be global and deferred, while those from adaptation projects are expected to be local and to some extent more immediate (World Bank 2009). Important adaptation options in agricultural sector include: crop diversification, mixed crop-livestock farming systems, using different crop varieties, changing planting and harvesting dates, and mixing less productive, drought-resistant varieties and high-yield water sensitive crops (Bradshaw et al. 2004).

The baseline survey indicate that CC projects implemented in Kenya during the past five years were mostly on adaptation as agreed by 60 % of the respondents, followed by projects in capacity building (23 %) and mitigation (17 %) respectively (Fig. 4.1). These findings are fairly rational given that adaptation is a way of trying to tolerate and live with the CC, while capacity building is empowering people in raising awareness, training and education and providing other capacity

Fig. 4.1 Projects addressing selected climate change thematic areas (n = 263)

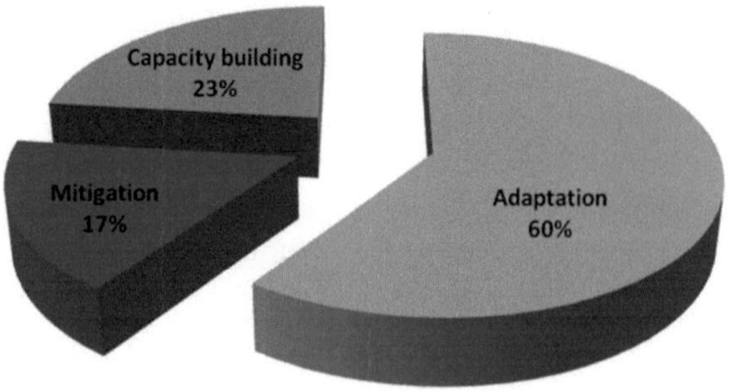

requirements to deal with and accommodate climate change (CC) scenarios. Some projects have emphasized in enhancing provision of climate information services, strengthening capacity of governments to facilitate adaptation to CC, building awareness and capacity among civil society; and to a lesser extent improving freshwater resources, pastoralism and human health (Kurukulasuriya and Rosenthal 2003).

Projects in Selected Productive Sectors

Kenya's productive sectors are the most sensitive ecosystems to climate change and variability (CCV). Some of these sectors were identified as agriculture, livestock, water, tourism, health, infrastructure, natural resources (the environment), and fisheries (Kpadonou et al. 2012, IPCC 2007; IFPRI 2007; World Bank 2007). By expert opinion and consensus, four most sensitive sectors to CC were identified for analysis of this research. The sectors are agriculture (crops), livestock, environment (natural resources), and water resources.

It was observed by 35.7 % of the respondents that agriculture sector accounted for most of climate change (CC) projects during the past five years in Kenya. Similarly, livestock, environment and water resources sectors accounted for 27.4, 19.8 and 17.1 % of the projects during the same period (Fig. 4.2). This finding clearly indicates that agriculture and livestock (63.1 %) accounted for the bulk of the CC projects in Kenya. One of the reasons underpinning this trend could be that agriculture and livestock sectors are more directly related to food security than any

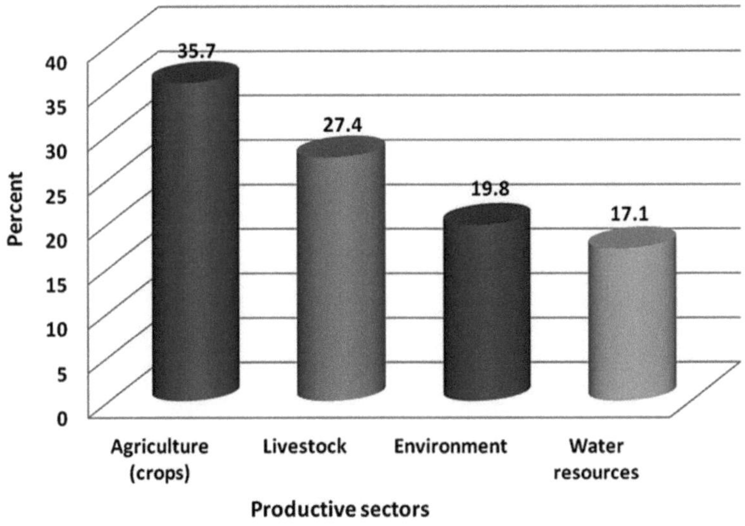

Fig. 4.2 Projects addressing selected productive sectors (n = 263)

other sector. Further, the effects of climate change and variability (CCV) are easily and immediately reflected on the production of crops and livestock commodities.

Adaptation Projects in Selected Productive Sectors

Given that most climate change (CC) projects in Kenya were implemented within the adaptation theme, it became apparent to reflect how the thematic projects were implemented and distributed in the selected productive sectors. This provided reflections on priorities areas in which investments on CC projects are made. It is shown that adaptation projects were mostly invested in agriculture sector accounting for 39.5 % of all adaptation projects implemented in Kenya during the past five years. This was followed by projects in livestock (27.4 %), environment (17.2 %) and water resources (15.9 %) (Fig. 4.3). Again, agriculture and livestock sectors (66.9 %) put together accounted for the bulk of the adaptation projects implemented in Kenya. The moderately high levels of investments in adaptation projects in agriculture and livestock are encouraging given that these two sectors are critical in their contribution to the Kenyan economy and food security. These investment levels need to be enhanced in these sectors given their vulnerability to climate change and variability (CCV) as well as their importance to food security and economic growth.

Mitigation Projects in Selected Productive Sectors

The survey analysis shows that mitigation projects have been going on in Kenya during the past five years. It is shown that the three sectors: agriculture, livestock and environment each accounted for about 29.5 % of the mitigation projects in

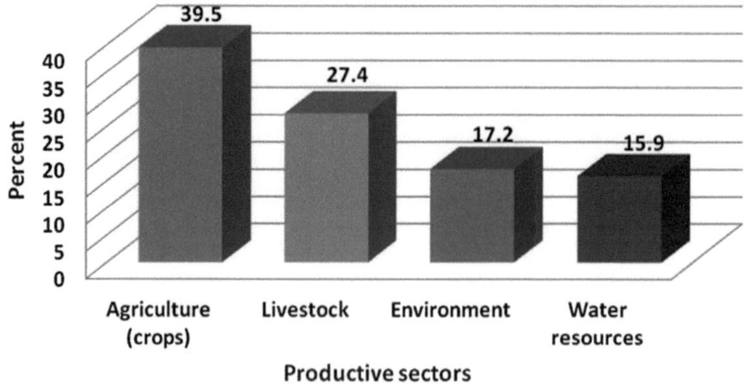

Fig. 4.3 Adaptation projects addressing selected productive sectors (n = 157)

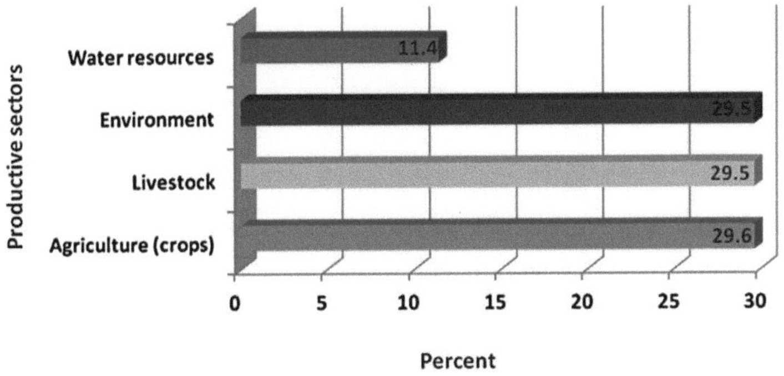

Fig. 4.4 Mitigation projects addressing selected productive sectors (n = 44)

Kenya during this period. In spite of the increases in frequency and severity of floods in Kenya, water resources accounted for a paltry of 11.4 % only of the mitigation projects in the country (Fig. 4.4). This may explain the massive destruction of property and lose of livelihoods reported every rainy season.

Notwithstanding, it is generally recognised that smallholder farmers can contribute substantially to climate change (CC) mitigation, but will need incentives to adapt mitigation practices. These incentives would include the selling of carbon credits, which unfortunately are limited by low returns to farmers, high transaction costs, and the need for farmers to invest in mitigation activities long before they receive payments. Designing agricultural investments and policies to provide up-front financing and longer term rewards for mitigation practices will help reach larger numbers of farmers than specialized mitigation interventions (Wollenberg et al. 2012).

It is instructive to note that potential for mitigation strategies is great and what is needed is a coordinating strategy to organise the generation and sharing of greenhouse gas data, and facilitate improved understanding of the potential for greenhouse gas emissions and removals from the CC sensitive sectors such as agriculture and forestry.

In Kenya mitigation activities have been practised on crop and soil management practices including sustainable agriculture land management, nutrient management (fertilisers), tillage and residue management, and agroforestry. Mitigation has also been practised on livestock and grazing land management that included grazing intensity—intensification and reduced herd sizes (productivity), and rangeland and pastureland management (Masiga 2012).

Capacity Building Projects in Selected Productive Sectors

It is observed (Fig. 4.5) that capacity building projects were mostly undertaken in agriculture sector that accounted for 30.6 % of all the projects in the Kenya. This was followed by livestock (25.8 %), water resources (24.2 %) and environment

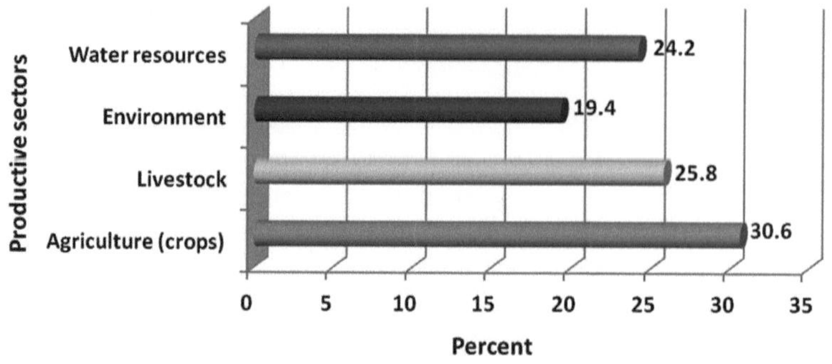

Fig. 4.5 Capacity building projects addressing selected productive sectors (n = 62)

(19.4 %) sectors. Up to 81 % of all the capacity building projects were undertaken in agriculture, livestock and water resources sectors.

One example of the capacity building project going on in Kenya is the 'building adaptation capacities for climate change (CC) through participatory research, training and outreach', which was initiated in 2010. This project is evaluating indigenous/traditional CC mitigating and adaptation strategies currently used by diverse Kenyan farming and pastoral communities and build capacity on CC adaptation strategies among various stakeholders (Lelo 2011).

Conclusions and Recommendations

Most projects undertaken in Kenya on climate change (CC) arena have been on adaptation, capacity building and mitigation areas, while majority of the CC projects undertaken were in agriculture and livestock sectors. Three sectors on agriculture, livestock and environment received an equal share of mitigation projects, while majority of the CC capacity building projects were implemented in agriculture, livestock and water resources sectors.

Given the importance of adaptation in tolerating effects/impacts of CC, it is recommended that more adaptation work be intensified in Kenya. One area to work on is to undertake policy review to provide enabling environment to conduct adaptation research for development. Capacity building should also be embraced to increase awareness, education and training, and tools and equipment for CC issues.

Acknowledgments The authors are grateful to International Development Research Centre (IDRC) for funding and Kenya Agricultural Research Institute (KARI) Director for logistical support. We would also like to acknowledge the steering committees of 27th Soil Science Society of East Africa (SSSEA) and 6th African Soil Science Society (ASSS) conference in Nakuru Kenya for allowing the presentation of this publication thereby gaining visibility. Any views expressed here are those of the authors.

References

African Conservation Centre (2012) The impact of the 2009 drought on wildlife, livestock and tourism: recommendations for ecosystem restoration. African Conservation Centre, Kenya

Beddington J (2010) Food security: contributions from science to a new and greener revolution. Philos Trans R Soc B 365:61–71

Bradshaw B, Dolan H, Smit B (2004) Farm-level adaptation to climatic variability and change: crop diversification in canadian prairies. Clim Change 67:119–141

Downing TE, Ringuis L, Hulme M, Waughray D (1997) Adapting to climate change in Africa. Mitig Adapt Strat Glob Change 2:19–44

GoK (2010) National climate change response strategy. Ministry of Environment and Natural Resources, Government of Kenya, Kenya

IFPRI (2007) Micro-level analysis of farmers' adaptation to climate change in Southern Africa. Discussion paper 00714. International Food Policy Research Institute, Washington, DC

IPCC (2007) Climate change 2007: impacts, adaptation and vulnerability. In: Parry ML, Canziani OF, Palutikof JP, van der Linden PJ, Hanson CE (eds) Contribution of working group II to the 4th assessment report of the intergovernmental panel on climate change, Annex I. Cambridge University Press, Cambridge, p 976

Knox J, Hess T, Dacache A, Wheeler T (2012) Climate change impacts on crop productivity in Africa and South Asia. Environ Res Lett 7:034032

Kpadonou RAB, Adégbola PY, Tovignan SD (2012) Local knowledge and adaptation and climate change in Ouémé Valley, Benin. Afr Crop Sci J 20(2):181–192

Kurukulasuriya P, Rosenthal S (2003) Climate change and agriculture: a review of impacts and adaptations. Paper no. 91 in climate change series, Agriculture and Rural Development Department and Environment Department, World Bank, Washington, DC

Lelo FK (2011) Building adaptation capacities for climate change in Kenya through participatory research, training and outreach, a project supported by the Rockefeller Foundation. Egerton University, Kenya. http://www.fao.org/climatechange/micca/75369/en/. Retrieved 24 Nov 2012

MacMillan Susan (2011). Predicted impacts of climate change on Kenya: definitely hotter–expect less productive cropping, more livestock herding. ILRI News. Fri 5 Aug 2011. http://www.ilri.org/ilrinews/index.php/archives/6879, 20 Nov 2012

Marland G, Boden TA, Andres RJ (2007) Global, regional, and national CO2 emissions. In: Trends: a compendium of data on global change. Carbon dioxide information analysis center, Oak Ridge National Laboratory, United States Department of Energy, Oak Ridge, Tenn, USA

Masiga M (2012) Economic and social evaluation of national GHG mitigation options for agricultural landscapes in East Africa. In: Proceedings of CCAFS/FAO expert workshop on NAMAs: national mitigation planning and implementation in agriculture, Rome, Italy, 16–17 July 2012

Molina M, Zaelke D, Sarmac KM, Andersen SO, Ramanathane V, Kaniaruf D (2009) Reducing abrupt climate change risk using the montreal protocol and other regulatory actions to complement cuts in CO2 emissions. Proc Natl Acad Sci 106(49):20616–20621

Müller C, Cramer W, Hare WL, Lotze-Campen H (2011) Climate change risks for African agriculture. Proc Natl Acad Sci USA 108:4313–4315

Mutimba S, Mayieko S, Olum P, Wanyama K (2010) Climate change vulnerability and adaptation preparedness in Kenya. Heinrich Böll Stiftung, East and Horn of Africa, Nairobi, 2010, p 30

Nanyingi MO, Kiama SG, Thumbi S, Muchemi GM (2012) Climate change vulnerability, adaptation and mitigation of livestock systems in Kenya. Wangari Maathai Institute for Environmental Studies and Peace, College of Agriculture and Veterinary Science, University of Nairobi, Nairobi, Kenya

Pan African Climate Justice Alliance (2009) PACJA statement to the 2nd meeting of the African high level expert panel on climate change, Addis Ababa, Ethiopia, 22 Oct 2009

Schlenker W, Lobell DB (2010) Robust negative impacts of climate change on African agriculture. Environ Res Lett 5:014010

Slingo JM, Challinor AJ, Hoskins BJ, Wheeler TR (2005) Introduction: food crops in a changing climate. Phil Trans R Soc B 360:1983–1989

UNFCCC (1997) Glossary of climate change acronyms. Retrieved 22 Nov 2012

Wollenberg E, Higman S, Seeberg-Elverfeldt C, Neely C, Tapio-Bistrom M-L Neufeldt H (2012) Helping smallholder farmers mitigate climate change. CCAFS policy brief 5. CCAFS, Copenhagen, Denmark

World Bank (2007) Overview on international trade and climate change: economic, legal and institutional perspectives. International Bank for Reconstruction and Development/World Bank, Washington, DC, p 26

World Bank (2009) The economics of adaptation to climate change: a final methodology report. World Bank, Washington, DC

Chapter 5
Seasonal Rainfall Variability and Drought Characterization: Case of Eastern Arid Region, Kenya

M. Oscar Kisaka, M. Mucheru-Muna, F. Ngetich, J. Mugwe, D. Mugendi and F. Mairura

Abstract Drier parts of Embu County, Eastern Kenya, endure persistent crop failure and declining agricultural productivity which have been attributed, in part, to prolonged dry-spells and erratic rainfall. Nonetheless, understanding spatial-temporal variability of rainfall especially at seasonal level, is an imperative facet to rain-fed agricultural productivity and natural resource management (NRM). This study evaluated the extent of seasonal rainfall variability and the drought characteristics as the first step of combating declining agricultural productivity in the region. Cumulative Departure Index (CDI), Rainfall Anomaly Index (RAI) and Coefficients-of-Variance (CV) and probabilistic statistics were utilized in the analyses of rainfall variability. Analyses showed 90 % chance of below cropping-threshold rainfall (500 mm) exceeding 213.5 mm (Machanga) and 258.1 mm (Embu) during SRs for one year return-period. Rainfall variability was found to be high in seasonal amounts (CV = 0.56 and 0.38) and in number of rainy-days (CV = 0.88 and 0.27) at Machang'a and Embu, respectively. Monthly rainfall variability was found to be equally high even during April (peak) and November (CV = 0.42 and 0.48 and 0.76 and 0.43) with high probabilities (0.40 and 0.67) of droughts exceeding 15 days in Embu and Machang'a, respectively. Dry-spell probabilities within growing months were high (81 %) and (60 %) in Machang'a

M. Oscar Kisaka (✉) · M. Mucheru-Muna
Department of Environmental Science, Kenyatta University, P.O. Box 43844-00100
Nairobi, Kenya
e-mail: soscarneyo@gmail.com; O.Masika@cgiar.org

F. Ngetich · J. Mugwe
Departments of Agricultural Resource Management, Kenyatta University,
P.O. Box 43844-00100, Nairobi, Kenya

D. Mugendi
Embu University College, P.O. Box 6-60100, Embu, Kenya

F. Mairura
Tropical Soil Biology and Fertility Institute of CIAT (TSBF-CIAT), P.O. Box 30677
Nairobi 00100, Kenya

W. Leal Filho et al. (eds.), *Adapting African Agriculture to Climate Change*,
Climate Change Management, DOI 10.1007/978-3-319-13000-2_5

and Embu respectively. To optimize yield in the area, use of soil-water conservation and supplementary irrigation, crop selection and timely accurate rainfall forecasting should be prioritized.

Keywords Cumulativc-dcparturc-indcx · Drought-probability · Rainfall-anomaly-index · Rainfall-variability

Introduction

Understanding spatio-temporal patterns in rainfall has been directly implicated to combating extreme poverty and hunger through agricultural enhancement and natural resource management (IPPC 2007). The amount of soil-water available to crops depends on onset, length and cessation of rainy season which influence the success or failure of a growing season (Ati et al. 2002). It's thus palpable that, climatic parameters and rainfall in particular are prime inputs of improving the socio-economic wellbeing of smallholder farmers. This is particularly important in Sub-Saharan Africa (SSA) where agricultural productivity is principally rain-fed yet highly variable (Jury 2002). Drier parts of Embu County, Eastern Kenya experience unpredictable rainfall patterns, persistent dry-spells/droughts coupled with high annual potential evapo-transpiration (2,000–2,300 mm year^{-1}) (Micheni et al. 2004). There is generally enough water on the total; however, it is poorly distributed over time (Kimani et al. 2003) with 25 % of the annual rain often falling within a couple of rainstorms, that crops suffer from water stress, often leading to complete crop failure (Meehl et al. 2007; Recha et al. (2012) noted that, most studies do not provide information on the much-needed character of within-season variability despite its implication on soil-water distribution and productivity. There has been continued interest in understanding seasonal rainfall patterns by evaluation of its variables including rainfall amount, rainy days, lengths of growing seasons and even dry-spell frequencies. Studies by Sivakumar (1991), Seleshi and Zanke (2004) and Tilahun (2006) noted high variations in annual and seasonal rainfall totals and rainy days in Ethiopia and Sudano-Sahelian regions. Studies on rainfall patterns in the region have been based principally on annual averages, thus missing on within-season rainfall characteristics (Barron et al. 2003). Nonetheless, understanding the average amount of rain per rainy day and the mean duration between successive rain events aids in understanding long-term variability and patterns (Akponikpè et al. 2008). Hitherto, the much-needed information on inter/intra seasonal variability of rainfall in the region is still inadequate despite its critical implication on soil-water distribution, water use efficiency (WUE), nutrient use efficiency (NUE) and final crop yield. To optimize agricultural productivity in the region, there was need to quantify rainfall variability at a local and seasonal level as a first step of combating extreme effects of persistent dry-spells/droughts and crop failure. Since rainfall, in particular, is the most critical factor determining rain-fed agriculture yet not homogeneous, knowledge of its statistical properties derived from long-term observation could be utilized in developing variability and drought mitigation strategies in the area.

Materials and Methods

The Study Area

The study areas is covered by agro-ecologies classified as lower midland 3, 4 and 5 (LM 3, LM 4 and LM 5), Upper midland 1, 2, 3 and 4 (UM 1, UM 2, UM 3 and UM 4), and Inner lowland 5 (IL 5) (Jaetzold et al. 2007) and lies at an altitude of approximately 500–1,800 m above mean sea level (Fig. 5.1).

It has an annual mean temperature ranging from 21.7 to 22.5 °C and average annual rainfall of 700–900 mm. It has a population density of 82 persons per km^2 with an average farm size less than 5.0 ha per household. The rainfall is bimodal with long rains (LR) from mid-March to June and short rains (SR) from late October to December hence two cropping seasons per year. The soils are predominantly Ferralsols and Acrisols (Jaetzold et al. 2007). Various agricultural-based studies have been carried out in the region hence the rationale behind its selection. According to Mugwe et al. (2009), the region has experienced drastic declines in its productivity potential rendering its populace resource poor. There is a secure tenure system on land ownership but underscore in productivity due to inadequate information on the rainfall patterns. The prime cropping activity is maize intercropped with beans though livestock keeping is equally dominant. Mbeere Sub-county represents a sub-humid climate region, with annual average rainfall of 781 mm while Embu is more humid with annual average rainfall above 1,210 mm (Table 5.1).

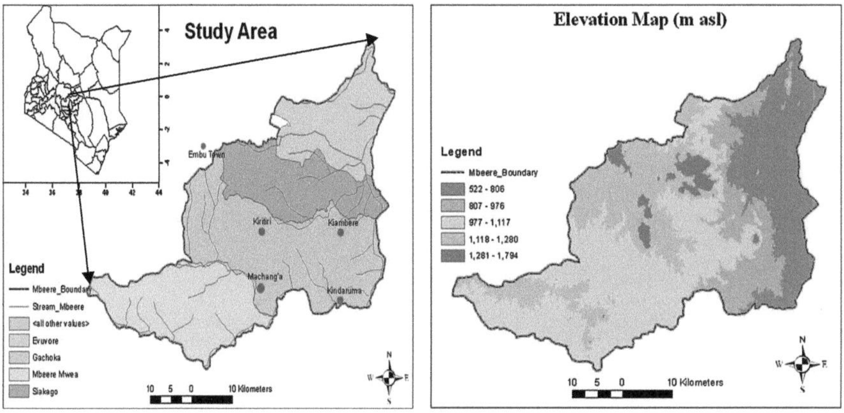

Fig. 5.1 Map showing the study area and its elevation with selected point gauged rainfall data; Machang'a and Embu, Kiritiri, Kindaruma and Kiambere

Table 5.1 Selected agro-climatic characteristics of the meteorological stations (Embu and Machang'a) used in the study

Station	Latitude	Longitude	Altitude	Period of record	Rainfall	Climate	Data
Embu	0°30′S	37°27′E	1409	13	1,210	humid	R
Machang'a	0°46′S	37°39′E	1,106	13	781	s-humid	R

Daily rainfall, and maximum/minimum temperature and solar radiation data were sourced from both the Kenya Meteorology Department and research sites with primary recording stations within Mbeere Sub-county. The choice of rainfall stations used relied on the agro-ecological zones, the percentage of missing data, [less than 10 % for a given year as required by the world meteorological organization (WMO)].

Data Analyses

Daily primary and secondary rainfall time series were captured into MS Excel spread-sheet where seasonal rainfall totals for both Short Rains (SR) and Long Rains (LR) [that is, March-April-May (MAM) and October-November-December (OND) respectively], annual average and number of rainy days were computed. Multiple imputations were utilized to fill in missing daily data through creation of several copies of datasets with different possible estimates. The multiple imputation method was preferred to single imputation and regression imputation as it appropriately adjusted the standard error for missing data yielding complete data sets for analysis (Enders 2010). Being a season-based analysis, the cumulative impact of rainfall amount was underpinned. A rainy day was considered to be any day that received more than 0.2 mm of rainfall. Daily rainfall data were captured into the *RAINBOW software* (Raes et al. 2006) for homogeneity testing based on cumulative deviations from the mean to check whether numerical values came from the same population. The cumulative deviations were then rescaled by dividing the initial and last values of the standard deviation by the sample standard deviation values; as in the Eq. (5.1) below;

$$S_k = \sum_{i=1}^{k} \left(X_i - \overline{X}\right) \qquad \text{when } k = 1, \ldots, n \tag{5.1}$$

where S_k is the Rescaled Cumulative Deviation (RCD), n represents the period of record for $K = 1$ and also when $K = 14$.

The maximum (Q) and the range (R) of the rescaled cumulative deviations from the mean were evaluated based on number of Nil Values, Non-Nil values, Mean and Standard deviations as well as K–S values (Eqs. (5.2) and (5.3)) to test homogeneity. Low values of Q and R would indicate that data was homogeneous.

$$Q = \max\left[{}^{S_k}/_S\right] \tag{5.2}$$

$$R = \max\left[{}^{S_k}/_S\right] - \min\left[{}^{S_k}/_S\right] \tag{5.3}$$

where Q is maximum (max) of S_K and R in the range of S_K and Min is Minimum.

The frequency analyses were based on lognormal probability distribution with $\log_{10}$ transformation using cumulative distribution function (CDF) for both LR and SR rainfall amounts. The Weibull method was used to estimate probabilities while the Maximum Likelihood Method (MOM) was utilized as a parameter estimation statistic. Homogeneous seasonal rainfall totals for both LRs and SRs was then subjected to trend and variability analyses based on Cumulative Departure Index (CDI) and Rainfall Anomaly Index (RAI) as described in Tilahun (2006). Trend analyses based on CDI utilized normalized arithmetic means for seasonal and annual rainfall for the period of record (14 years) using Eq. (5.4).

$$CDI = (r - R)/S \tag{5.4}$$

where r is actual rainfall (seasonal or annual), R is the mean rainfall of the total length of period recorded, S is the standard deviation of the total length of period of record.

Seasonal Variability was computed in tandem with annual averages for both positive (Eq. 5.5) and negative (Eq. 5.6) anomalies using RAI.

$$RAI = +3\left(\frac{RF - M_{RF}}{M_{H10} - M_{RF}}\right) \tag{5.5}$$

$$RAI = -3\left(\frac{RF - M_{RF}}{M_{L10} - M_{RF}}\right) \tag{5.6}$$

where M_{RF} is mean of the total Length of record, M_{H10} is mean of 10 highest values of rainfall of the period of record, M_{L10} is the lowest 10 values of rainfall of the period of record.

The Coefficient of Variance (CV) statistics were utilized to test the level of mean variations in LR and SR seasonal rainfall, number of rainy days (RD) and Rainfall Amounts (RA) and t-test statistic to evaluate the significance of variation.

A dry day was taken as a day that received less than 0.2 mm rainfall. A dry spell was considered as sequence of dry days bracketed by wet days on both sides (Kumar and Rao 2005). The method for frequency analysis of dry spells was adapted from Belachew (2000) as follows: in the Y years of records, the number of times (i) that a dry spell of duration (t) days occurs was counted on a monthly basis. Then the number of times (I) that a dry spell of duration longer than or equal to *t* occurs was computed through accumulation. The consecutive dry days (1 d, 2 d, 3 d, …) were prepared from historical data. The probabilities of occurrence of consecutive dry days were estimated by taking into account the number of days in a given month *n*. The total possible number of days, *N*, for that month over the analysis period was computed as, $N = n * Y$. Subsequently the probability *p* that a dry spell may be equal to or longer than *t* days was given by Eq. (5.7): The probability q that a dry spell not longer than t does not occur at a certain day in a growing season was computed by Eq. (5.8); and probability Q that a dry spell longer than t days will occur in a growing season was calculated by Eq. (5.9) and probability that a dry spell exceeding t days would occur within a growing season was computed by Eq. (5.10) as shown below:

$$P = {}^{1}\!/_{N} \tag{5.7}$$

$$q = (1 - p) = \left[1 - \frac{1}{N}\right] \tag{5.8}$$

$$Q = \left[1 - \frac{1}{N}\right]^{n} \tag{5.9}$$

$$p = (1 - Q) = 1 - \left[1 - \frac{1}{N}\right]^{n} \tag{5.10}$$

Results and Discussion

Homogeneity Testing

Homogeneity analyses had no NIL-values (values below threshold) but 100 % Non-Nil values (above threshold) showing high homogeneity. The standard deviations (SD) of the normalized means for both LR and SR seasons were low (SD = 0.2, and SD = 0.9 and 0.1) at Machang'a and Embu, respectively, indicating restriction of variations rescaled cumulative deviations (RCD), thus high homogeneity (Table 5.2).

Table 5.2 Mean, standard deviation and R2 values for Embu and Machang'a rainfall dailies for the period between 2001 and 2013

Season	Transformation	Nil values		Mean		SD		R^2 (%)	
		Mac	Emb	Mac	Emb	Mac	Emb	Mac	Emb
LR	Log_{10}	0	0	2.4	3.2	0.2	0.9	96	94
SR	Log_{10}	0	0	2.6	2.6	0.2	0.1	94	92

Mac Machang'a, *Emb* Embu and *SD* Standard deviation; *LR* Long rains and *SR* Short rains

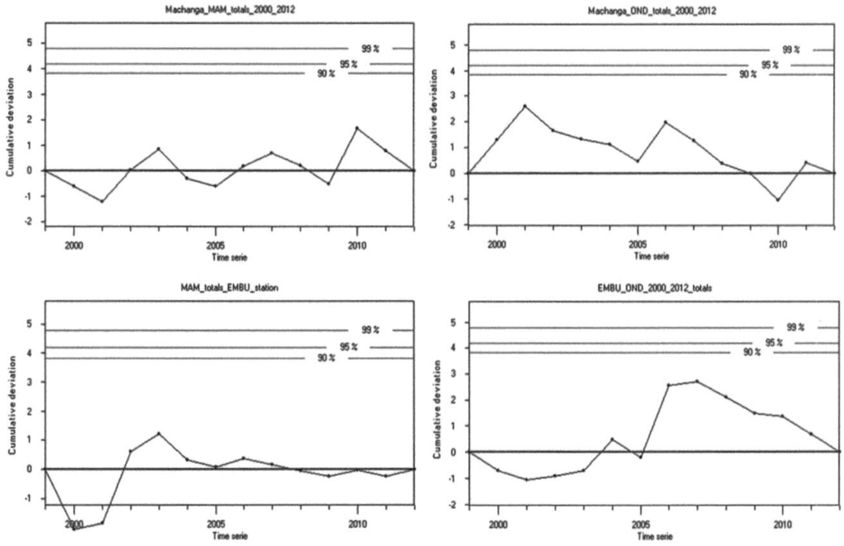

Fig. 5.2 Rescaled cumulative deviations for LR (MAM) and SR (OND) seasonal rainfall for the period between 2000 and 2013. **a** Machang'a LRs. **b** Machang'a SRs. **c** Embu LRs and **d** Embu SRs

A plot of homogeneity showed deviations from the zero mark of the RCDs not crossing probability lines thus, homogeneity was accepted at 99 % probabilities (Fig. 5.2). There was a normal distribution of the sampled-temporal rainfall data with high goodness-of-fit (R^2 = 92–96 %) of the selected distribution showing continuity of the data from mother primary data thus high homogeneity (Raes et al. 2006). Kolmogorov Smirnov values (one sided sample K–S test) showed K–S values (0.15–0.23) consistently lower than the K–S table value (0.302) for $n = 14$ at $\alpha = 0.005$ probability indicating that an exponential, continuous distribution of the studied datasets was statistically acceptable, based on the empirical cumulative distribution function (ECDF) derived from the largest vertical difference between the extracted (observed k-s value) and the table value (Table 5.3) (Botha et al. 2007; Mzezewa 2010; MATLAB Central 2013).

Frequency analyses of meteorological data require that the time series be homogenous in order to gain in-depth and representative understanding of the trends over time (Raes et al. 2006). Often, non-homogeneity and lack of exponential distributions between datasets indicate gradual changes in the natural environment (thus trigger variability) which corresponds to changes in agricultural production (Huff and Changnon 1973; Bayazit 1981).

Table 5.3 Homogeneity test for Embu and Machang'a rainfall dailies for the period between 2000 and 2013

Month	Transformation	(K–S)		N		K–S:T. value	
		Mac	Emb	Mac	Emb	Mac	Emb
LR	Log10	0.1479	0.2330	14	14	0.302*	0.302*
SR	Log10	0.1900	0.1722	14	14	0.302*	0.302*

K–S Kolmogorov Smirnov Test; *Table V* Table Value; (0.302* exponential distribution applies and accepted) *Mac* Machang'a, *Emb* Embu

Rainfall Seasonality Patterns

Results showed that there was at least 90 % chance of rainfall exceeding 172.2 and 213.5 mm during LRs in Machang'a and Embu, respectively, within a return period of about 1 year (Table 5.4). Nonetheless, there were observably low probabilities (10 %) that rains would exceed 449.8 and 763.0 mm during LR seasons in Machang'a and Embu, respectively for a 10-year return period (Table 5.4). Seasonal rainfall averages were equally low, especially in Machang'a (314.9 and 438.7 mm).

A study by Mzezewa (2010) established that seasonal rainfall amount greater than 450 mm is indicative of a successful growing season; and described it as a threshold rainfall amount. During this study, the probabilities that seasonal rainfall would exceed this threshold were quite low (at most 30 % for a return period of 3.33 years). Embu, being much wetter, would probably (50 %) receive above threshold rainfall amount (506.8 mm) after every 2 years (Table 5.4). Studies agreeing with these findings include Mzezewa (2010) who studied the semi-arid Ecotope of Limpopo South Africa. Mzezewa (2010) observed 47 % chance of seasonal-rainfall exceeding 580 mm but 0 % (no increase) of exceeding total annual rainfall for a 5-year return period.

Table 5.4 Probability of rainfall exceedance and return-periods for the LRs and SRs at Machang'a and Embu

Probability of exceedance (%)	Return period (year)	Magnitude of anticipated rainfall (mm)			
		LR		SR	
		Machang'a	Embu	Machang'a	Embu
10	10	449.8	994.7	763.0	628.8
20	5	381.4	788.9	613.1	541.2
30	3.33	338.7	667.5	523.7	485.7
40	2.50	306.0	578.8	457.7	442.9
50	2	278.2	506.8	403.6	406.3
60	1.67	253.2	443.5	356.0	372.8
70	1.43	222.8	384.5	311.1	339.9
80	1.25	203.1	325.4	265.7	305.0
90	1.11	172.2	258.1	213.5	262.5

LR Long Rains March-May-June and *SR* Short Rains October-November-December and (*mm*) millimetres

Trend of the Rainfall Events

Rainfall trends of the studied period showed that SRs and LRs in Machang'a were persistently above and below average respectively. In the former season, high rainfall was received between 2006 and 2007 (CDI = +2.5) while in the latter season, above average rainfall amounts were experienced only twice (2001 with CDI = +1 and in 2012 with CDI = +2) (Fig. 5.3). During the same period, high fluctuations in seasonal rainfall amounts were recorded in Embu. A general decline in LRs and annual averages was observed from 2003 (CDI = 1.5 and 0.5) to 2010 (CDI = −2 and −1), respectively (Fig. 5.4).

The high variability trends in seasonal and annual rainfall amounts observed in this study corroborate findings by Nicholson (2001), Hulme (2001) and Dai et al. (2004). In this study, the decade between 2000 and 2013 experienced marked increase in SRs and a decrease in LRs. Nicholson and Hulme (2001) attributed the decrease in LRs to the desiccation of the March-to-August rains in Sub-Saharan Africa (SSA). A study by Tilahun (2006) based on the cumulative departure index established that parts of Northern and Central Ethiopia persistently received below average rainfall for the rains received between February and August since 1970. While studying vegetation dynamics based on the normalized difference vegetation index (NDVI), Tucker and Anyamba (2005) noted persistent droughts and unpredictable rainfall patterns marked by reduction in the NVDI values during LRs for periods approaching the 21st century. On the other hand, it was apparent that SRs recorded consistent above-average trends during this study; indicating possibilities of a reliable growing season especially for the drier Machang'a region. In tandem with this observation, findings by Hansen and Indeje (2004) and Amissah-Arthur et al. (2002) observed that SRs constituted the main growing season in the drier parts of SSA and Great Horne of Africa for crops such as maize, sorghum, green grams and finger millet.

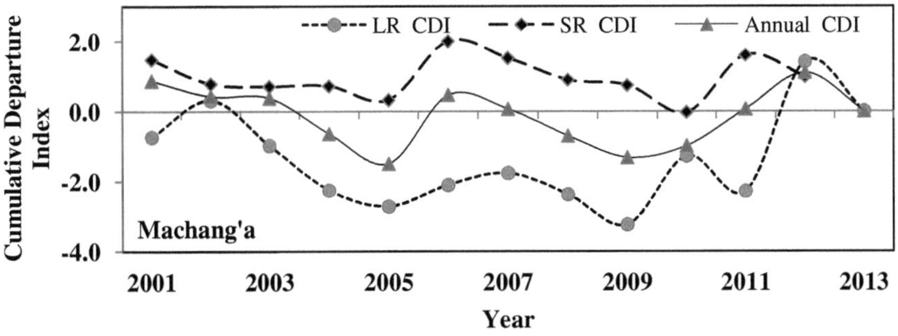

Fig. 5.3 Trend analyses based on cumulative departure index (*CDI*) for Machang'a rainfall station. (Fluctuations around, above and below the CDI zero mark corresponds to the deviations from the average rainfall for the period between 2001 and 2013)

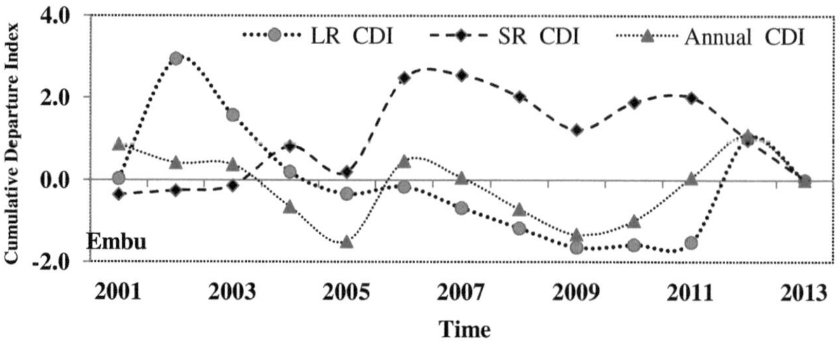

Fig. 5.4 Trend analyses based on cumulative departure index (*CDI*) for Embu rainfall station. (Fluctuations around, above and below the CDI zero mark corresponds to the deviations from the average rainfall for the period between 2001 and 2013)

Variability and Anomalies in Seasonal Rainfall Amount

There was high inter-seasonal variability and temporal anomalies in rainfall between 2001 and 2013. Results showed neither station nor season with persistent near average (RAI = 0) rainfall. The wettest LRs were recorded in 2010 (RAI = +4) while wettest SRs were recorded in 2001 (RAI = +4), 2006 (RAI = +3.8) and 2011 (RAI = +4) at Machang'a (Fig. 5.5). On overall, Machang'a recorded more negative anomalies in rainfall amount received compared to Embu.

In Embu, the highest positive anomalies (+5.0) were recorded in 2002, 2005 and 2007 during LRs (Fig. 5.6). There was an observable return period of positive anomalies after every two or three (years), e.g. 2002 (+5), 2005 (+5.0), and 2007 (+5.0) during LRs. No such distinct trend was observed in SRs (Fig. 4.5). Noticeably, Embu appeared to be receiving more near average rainfall during SRs (2002, 2003, 2007 and 2011) contrary to the trends observed in Machang'a.

An intra-station-seasonal comparison showed that SRs in Embu were less variable but more drier compared to LR seasons. Conversely, SRs in Machang'a were wetter than SRs in Embu but more variable in the former. Trends of more variable

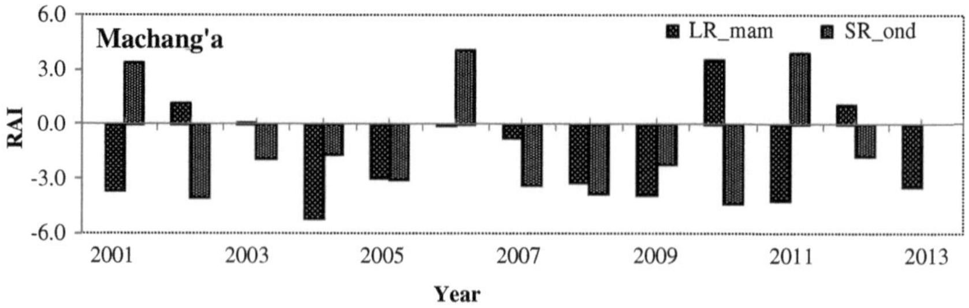

Fig. 5.5 Decadal rainfall anomaly index for both LR_MAM and SR_OND in Machang'a; Rainfall Anomaly Index (*RAI*)

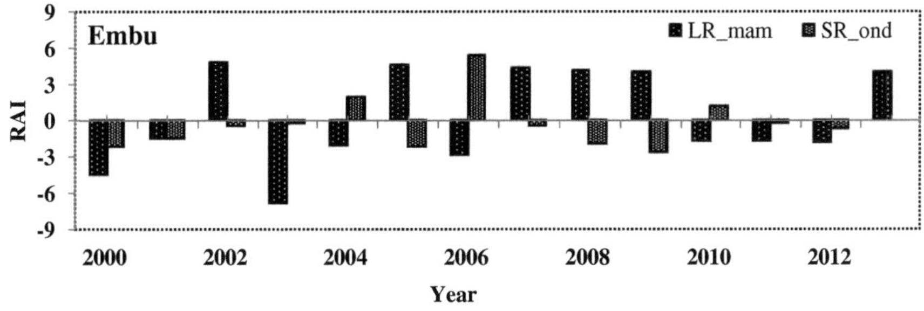

Fig. 5.6 Decadal rainfall anomaly index for both LRs and SRs in Embu site for the period between 2000 and 2013

but wetter SRs and less variable but drier LRs have been recorded in other studies such as Cohen and Lewis (1987); who documented the national drought of 1984 in Kenya, Shisanya (1990) and Recha et al. (2012). For instance, the failure of the LRs in 1984 prompted the Kenyan government to launch a national relief fund among other responses (Shisanya 1990). Akponikpè et al. (2008) concurred with this findings by reporting high variability (CV = 57 %) in temporal rainfall at annual (mono-modal rainfall between February and September), monthly and daily timescales in the Sahel region. High variability (attributed often to La Nina, El Nino and Sea Surface Temperatures) could occasion rainfall failures leading to declines in total seasonal rainfall in the study area. According to Shisanya (1990), La Nina events significantly contributed to the occurrence of persistent droughts and unpredictable weather patterns during LRs in Kenya. In contrast, El Nino events (of 1997 and 1998) have been cited as the key inputs of the positive anomalies in SR seasonal rainfall in the ASALs of Eastern Kenya (Anyamba et al. 2001; Amissah-Arthur et al. 2002).

Variations in Rainfall Amounts and Number of Rainy Days

On average, the total amount of rainfall received in Machang'a and Embu were below 900 and 1,400 mm per annum, respectively. Yet LRs contributed 314.9 and 586.3 mm while SRs contributed 438.7 and 479.1 mm (Table 5.5) translating to a total of 754 and 1,084 mm of seasonal rainfall in Machang'a and Embu, respectively (Table 5.5). These account for close to 90 % of total rainfall received annually; implying that smaller proportions of rainy days supplied much of the total amounts of rainfall received in the region. Generally, a Coefficient of Variation (CV) greater than 30 % indicates large variability in rainfall amounts and distributional patterns (Araya and Stroosnijder 2011). In Machang'a, rainfall amounts during LRs were highly variable (CV = 0.41) than those in Embu (CV = 0.36). This variability is simultaneously replicated in the CVs of rainy days (0.26 and 0.09),

Table 5.5 Variability analyses: coefficient of variations in seasonal rainfall amounts and number of rainy days for Machang'a and Embu for the period between 2000 and 2013

	LR				SR				M. variations
Station	RA	CV	RD	CV	RA	CV	RD	CV	T-test values
Machang'a	314.9	0.41	24	0.26	438.7	0.56	53	0.88	0.111
Embu	586.3	0.36	46	0.09	497.1	0.38	40	0.27	0.035*

MAM March-May-June and *OND* October-November-December and; *RA* Rainfall amount in (mm), *RD* Rainy days, *CV* Coefficient of variation and *M.variations* Mean variations; *Significant at 0.05 level

during the same season in the respective stations. Nonetheless, there exists a significant differences ($p = 0.035$ at probabilities of 0.05) in seasonal rainfall amounts in Embu but not in Machang'a ($p = 0.111$ at probabilities of 0.05) (Table 5.5).

It is evident that rainfall variability in both agro-ecological zones is markedly high. Analyses based on RAI indicated high variability in SR rainfall amounts in the two station (Figs. 5.5 and 5.6); which is further affirmed by high CV of SR rainfall amounts in the two stations (CV 0.56 in Machang'a and CV = 0.38 in Embu) (Table 5.5). In terms of variability in rainy days, SR recorded highest variability; probably an indicator of high rainfall variability in SSA during SR seasons. A study by Barron et al. (2003) reported similar findings in a station in Machakos; of Kenya which recorded variability in number of rainy days as CV = 53 and 45 % during SR and LR seasons respectively. Lack of notable significance in intra seasonal rainfall amounts in the drier parts of Kenya (represented by Machang'a in this study) was also reported by Recha et al. (2012). Regionally, findings of Seleshi and Zanke (2004) further showed that annual and seasonal rainfall (*Kiremt* and *Belg seasons*) in Ethiopia were highly variable with CV values ranging between 0.10 and 0.50.

Monthly Variations in Seasonal Rainfall Amounts and Number of Rainy Days

Understanding dynamics of rainfall amount variability at a season's monthly level and in number of rainy days can guide on the choice of planting time, crop variety as well as understanding of variations in onset, duration and cessation of seasonal rainfall. During this study, results showed that rainfall amounts received within seasonal months (March-April-May; LRs and October-November-December; SRs) were highly variable (all with CV > 0.3). Notably, coefficient of variation in Rainfall Amounts (CV-RA) were quite high during the months of March (CV-RA = 0.98) and December (CV-RA = 0.86) in Machang'a and CV-RA = 0.61 (March) and CV-RA = 0.97 (December) in Embu (Table 5.6). Least variability in CV-RA were recorded in the months of April (CV-RA = 0.42) and November (CV-RA = 0.43) in Machang'a and Embu, respectively (Table 5.6). Variability in the number of rainy days (CV-RD) was equally high in the two study stations. For

Table 5.6 Variability in seasonal months: coefficient of variation in rainfall amounts and rainy days for Machang'a and Embu for the period between 2000 and 2013

	Mar	April	May	Oct	Nov	Dec
Machang'a						
RA (mm)	85.5	160.2	69.2	98.9	267.9	72.0
CV-RA	0.98	0.42	0.69	0.80	0.77	0.86
RD	8	11	5	14	29	10
CV-RD	0.61	0.22	0.61	0.35	0.23	0.34
Embu						
RA (mm)	110.1	300.8	175.6	175.1	250.3	71.8
CV-RA	0.61	0.48	0.54	0.66	0.43	0.97
RD	20	14	12	10	13	17
CV-RD	0.47	0.27	0.27	0.59	0.25	0.83

RA (mm) Rainfall amount in millimetres; *CV-RA* Coefficient of variation in rainfall amounts, *RD* Number of rainy days; *CV-RD* Coefficient of variation in rainy days

instance, March (CV-RD = 0.61 and CV-RD = 0.47) and December (CV-RD = 0.34 and CV-RD = 83) had the highest variability in the number of rainy days in Machang'a and Embu, respectively (Table 5.6).

Generally, onset months (March and October) and cessation months (May and December) received highly variable rainfall amounts compared to mid months. Machang'a, though; being more of an arid region, it generally recorded lower variability in number of rainy days during SR seasonal months compared to those recorded at Embu during the same season, evidence of reduced variability and wetting of SRs in the region. Evidently, the amount of rainfall and number of rainy days received in the past decade at Machang'a have been more consistent (temporally) in April and November but highly unpredictable in March (basis of onset) and December (cessation). This significantly affects the cropping calendar in rain-fed agricultural productivity of the region. It has been shown that a CV > 30 % indicated large variability in rainfall amounts and distribution patterns (Araya and Stroosnijder 2011). By comparing the coefficient of variation of rainfall amounts (average CV-RA = 0.75) and that of rainy days (Average CV-RD = 0.39) at Machang'a and (CV-RA = 0.61; CV-RD = 0.45) at Embu, it is evident that there is high variability in amounts and days of rainfall received in the past decade as their variability exceeded 30 %.

Nonetheless, lower values of CV-RD indicated that variations in rainy days have been fairly consistent compared to variations in rainfall amounts received. Notably, there seems to be simultaneous variability in the May rainfall amounts (CV-RA = 0.69) in relation to May rainy days (RD-CV = 0.61) at Machang'a but no clear trend at Embu would be established. This implies that variations in rainy days have been fairly proportional to the rainfall amount received in the month of May in Machang'a than Embu. However, its importance to the cropping calendar may not be quite significant because May is a cessation month. On the other hand, highly contrasting variability were observed for the month of December (CV-RA = 0.86 and CV-RD = 0.34) at Machang'a and April (CV-RA = 0.61 and CV-RD = 0.30) at Embu.

It would also appear that Machang'a receives more rainfall during SR season with November alone accounting for 60 % of total rainfall amount received while April accounts for 51 % of the LR rainfall in Machang'a. Conversely, Embu receives more rainfall during LRs with April accounting for about 52 % of total rainfall received. These findings at Machang'a corroborate those of Barron et al. (2003) and Amissah-Arthur et al. (2002) which demonstrated that parts of Eastern Kenya receive more SR than LR rainfall amounts. Mzezewa et al. (2010) also reported high coefficient of variation for annual (315 %) and seasonal (50–114 %) rainfall in semi-arid Ecotope, north-east of South Africa. Also, Sivakumar (1991) found that annual rainfall in the Sudano-Sahelian zone of West Africa is less variable than monthly rainfall.

Generally, SRs in (Machang'a) and LRs in (Embu) rainfall amount and rainy days are fairly spread through the season, potentially reducing the impact of within-season variability. Additionally, the rainfall amounts received in May and December (cessation) is little and might not be sufficient to buffer crops from agricultural drought, especially in Mbeere South (Machang'a) where soils are predominantly sandy loam and shallow (Acrisols, Ferralsols, and Cambisols) (Jaetzold et al. 2007). Also, the first and last months (of both seasons) are characterized by high CV for rainfall amount and rainy days. Similar findings are reported in Sivakumar (1991) in which onset (May) and cessation (October) months in Sudano-Sahelian zone are characterized by variations of over 100 %.

Probability and Frequency of a Dry-Spells and Implications on Crop Productivity

Dry-spells during cropping months are quite common that often trigger reduced harvests or even complete crop failures, especially in the drier arid parts of Eastern Kenya. Results showed that in Machang'a (AEZ 4 and 5) and Embu's (AEZ 1 and 2), the probability of occurrence of dry-spells of various durations varied from month to month of the growing season. Observably, lowest probabilities of dry-spells occurrence of all durations would be in April (LRs) and November (SRs) in both stations. High probabilities of dry-spells were in March (0.72 and 0.55) and December (0.8 and 0.6) in Machang'a and Embu respectively. The probability of having a dry-spell increased with shorter periods (for instance, more chance of having a 3 than a 10 or 21 day dry-spell) (Fig. 5.7). Probabilities of a 15-day dry-spell were relatively lower (0.4–0.6) in both stations. Similarly, the probabilities of experiencing a 21 day dry spell were 0.3 and 0.4 for Embu and Machang'a, respectively (Fig. 5.7).

On the other hand, the probabilities that dry spells would exceed these day-durations were equally high (Fig. 5.8). There was 70 % chance that dry spells

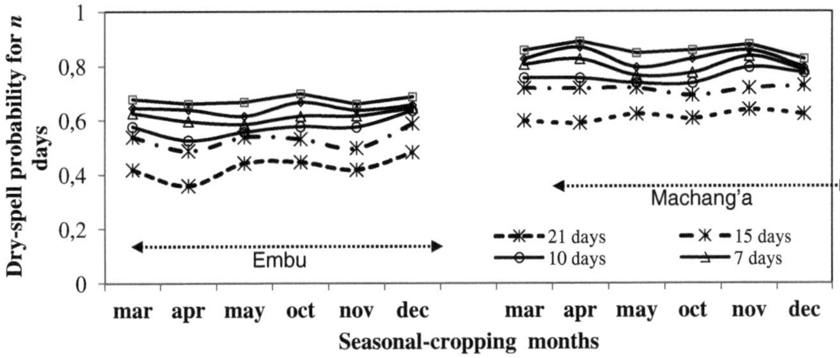

Fig. 5.7 Probability of a dry-spell of length ≥ n days, for n = 3, 5, 7, 15, 21, in each seasonal-cropping month, calculated using the raw rainfall data from 2000 to 2013 for stations in Machang'a and Embu

would exceed 15 days in Machang'a and 50 % in Embu (Fig. 5.8). It was also observed that April had high chances ($p = 0.85$) of its dry-spells exceeding 7 days in Machang'a while December recorded highest chances ($P = 0.6$) of its dry-spells exceeding 10 days in Embu; than any other month (Fig. 5.8).

Rainfall being a prime input and requirement for plant life in rain-fed agriculture, the occurrence of dry-spells has particular relevance to rain-fed agricultural productivity (Belachew 2000; Rockstrom et al. 2002). It was observed that lowest probabilities of occurrence of dry-spells of all durations were recorded in the month of April (during LRs) and November (during SRs). The occurrence of dry-spells of all durations decreased from April towards May (LR) and November towards December (SRs). Indeed, the months of April and December coincides with the peak of rainfall amounts for both SR and LR growing seasons in the region

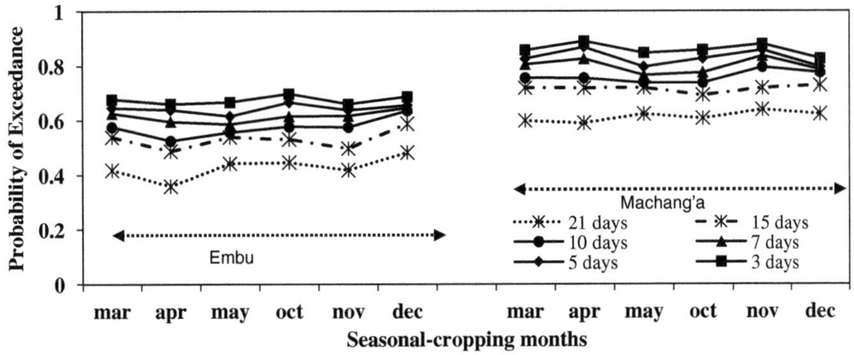

Fig. 5.8 Probability of dry-spells exceeding the *n* (3, 5, 7, 10, 15 *and* 21) days for each seasonal month calculated using the raw rainfall data from 2000 to 2013 for stations in Machang'a and Embu

respectively (Kosgei 2008; Recha et al. 2012). This trend is in line with works reported by several studies in SSA, including Kosgei (2008), Aghajani (2007) in Iran and Sivakumar (1992) in East Africa. Dry spells during SR season in Makindu and Katumani stations, lower eastern parts of Kenya had similar trends of high probabilities (avefinger milletng 88 %) in October. High probabilities of dry-spells occurring and exceeding the same durations show the high risks and vulnerability that rain-fed smallholder farmers are predisposed to in the study area. Often, prolonged dry-spells are accompanied by poor distribution and low soil moisture for the plant growth during the growing season. General high probabilities of persistent dry-spells in SSA have been reported by Hulme (2001), Dai et al. (2004) and Mzezewa (2010). Arguably, persistence of intermediate warming scenarios in parts of equatorial East Africa (Hulme 2001; Mzezewa 2010) may trigger increased dry-spells in months of May-August and January-March; further evidenced by the high probabilities of dry spells exceeding *n*th length days. Prolonged dry spells during cropping seasons directly impacts on the performance of crop production. For instance, high evaporative demand indicated by high aridity index ($P > 0.52$) in the drier parts of Eastern of Kenya implies that rain-water is not available for crop use and cannot meet the evaporative demands (Kimani et al. 2003). Thus, deficit is likely to prevail throughout the rain seasons as observed in other SSA regions (Li et al. 2003). Run-off collection and general confinement of rain-water within the crop's rooting zone could enhance rain-water use efficiency as demonstrated by Botha et al. (2007).

In most arid and semi-arid regions, soil moisture availability is primarily dictated by the extent and persistency of dry spells. It is thus essential to match the crop phenology with dry spell lengths based days after sowing to meet the crop water demands during the sensitive stages of crop growth (Sivakumar 1992). Knowledge of lengths of dry spells and the probability of their occurrence can also aid in planning for supplementary risk aversion strategies through prediction of high water demand spells. Information on lengths of dry spells also guides on the choice of crop types and varieties (Mzezewa 2010). For instance, probabilities of having dry spells exceeding 15 days is relatively low (23 and 15 % for Machang'a and Embu, respectively) during both SR and LR seasons. In this regard, the choice of crop variety and type should be based on the degree of its tolerance to drought (Sivakumar 1992; Mzezewa 2010). Most studies (including; Sivakumar 1992 and Belachew 2000) however indicate that decisions can be optimized if the probability of dry spells is computed after successful (effective) planting dates.

Conclusion and Recommendations

Decadal rainfall trends showed that both LRs and average annual rainfall have decreased in the past 13 years in both Embu and Machang'a. Machang'a appeared to have experienced pronounced declines in rainfall amounts especially those received during LRs. Nonetheless, rainfall amount during SRs markedly increased in both stations, with high amount gains established in Machang'a. Evidently, probabilities that seasonal rainfall amounts would exceed the threshold for cropping (500–800 mm) were quite low (10 %) in both stations. The amount of rainfall received during LRs and SRs varied significantly in Embu (t-test = 0.001 at $p < 0.05$) but not in Machang'a (t-test = 0.111, at $p < 0.05$). There was evidence of increasing rainfall variability for AEZ 1&2 (Embu) towards AEZ 4&5 (Machang'a) to as high as 88 % in CV. Probabilities that these AEZs would experience dry-spells exceeding 15 days during a cropping season were equally high, 46 % in Embu and 87 % in Machang'a. This replicates high chances that soil moisture could be lost by evaporation bearing in mind the high chances (81 %) the same dry-spells exceeding 15 days could reoccur during the cropping season. On the other hand, Kriging technique was identified as the most appropriate Geostatistical and deterministic interpolation techniques that can be used in spatial and temporal rainfall data reconstruction in the region. High rainfall variability and chances of prolonged dry spells established in this study demands that farmers ought to keenly select crop varieties and types that are more drought resistant (sorghum and millet) other than maize especially in the drier parts of Embu county (Machang'a). There is need for establishing further precise, timely weather forecasting mechanisms and communication systems to guide on seasonal farming.

Acknowledgments Special thanks are extended to RUFORUM, and the Principal Investigator Dr. Monicah Mucheru Muna, other participating project scientists; Prof. Daniel Mugendi, Dr. Jayne Mugwe, Mr. Felix Ngetich and Mr. Francis Mairura for their academic and fiscal support.

References

Aghajani GH (2007) Agronomical Analysis of the characteristics of the precipitation (Case study: Sazevar, Iran). Pakistan J Biol Sci 10(8):1353–1358

Akponikpè PBI, Michels K, Bielders CL (2008) Integrated nutrient management of pearl millet in the Sahel using combined application of cattle manure, crop residues and mineral fertilizer. Exp Agric 46:333–334

Amissah-Arthur A, Jagtap S, Rosen-Zweig C (2002) Spatio-temporal effects of *El Niño* events on rainfall and maize yield in Kenya. Int J Climatol 22:1849–1860

Anyamba A, Tucker CJ, Eastman JR (2001) NDVI anomaly pattern over Africa during the 1997/98 ENSO warm event. Int J Remote Sens 24:2055–2067

Anyamba A, Tucker CJ (2005) Analysis of Sahelian vegetation dynamics using NOAA-AVHRR NDVI data from 1981–2003. J Arid Environ 63:596–614

Araya A, Stroosnijder L (2011) Assessing drought risk and irrigation needs in Northern Ethiopia. Agric Water Manag 151:425–436

Ati OF, Stigter CJ, Oladipo EO (2002) A comparison of methods to determine the onset of the growing season in Northern Nigeria. Int J Climatol 22:731–742

Barron J, Rockstrom J, Gichuki F, Hatibu N (2003) Dry spell analysis and maize yields for two semi-arid locations in East Africa. Agric Meteorol 17:23–37

Bayazit M (1981) Hidrolojide İstatistik Yöntemler. İstanbul Teknik Üniversitesi, İstanbul, Yayın No: 1197, 223 s

Belachew A (2000) Dry-spell analysis for studying the sustainability of rain-fed agriculture in Ethiopia: the case of the Arba minch area. In: International commission on irrigation and drainage (ICID). Institute for the Semi-Arid Tropics, Direction de la meteorology nationale du Niger, Addis Ababa, Ethiopia, 116 p

Botha JJ, Anderson JJ, Groenewald DC, Mdibe N, Baiphethi MN, Nhlabatsi NN, Zere TB (2007) On-farm application of in-field rainwater harvesting techniques on small plots in the central region of South Africa. The Water Research Commission Report, South Africa

Cohen JM, Lewis DB (1987) Role of Government in Combatting Food Shortages: lessons from Kenya 1984/85. In: Glants MH (ed) Drought and Hunger in Africa, 269-2%. Cambridge University Press, Cambridge

Dai AG, Lamb PJ, Trenberth KE, Hulme M, Jones PD, Xie PP (2004) The recent Sahel drought is real. Int J Climatol 24:1323–1331

Enders CK (2010) Applied missing data analysis. The Guilford Press, New York. ISBN 978-1-60623-639-0

Graef F, Haigis J (2001) Spatial and temporal rainfall variability in the Sahel and its effects on farmers' management strategies. J Arid Environ 48:221–231

Hansen JW, Indeje M (2004) Linking dynamic seasonal climate forecasts with crop simulation for maize yield prediction in semi-arid Kenya. Agric For Meteorol 125:143–157

Huff FA, Changnon SA Jr (1973) Precipitation Modification by Major Urban Areas. Bull Amer Meteor Soc 54:1220–1232

Hulme M (2001) Climatic perspectives on SSA desiccation: 1973–1998. Glob Environ Change 11:19–29

IPCC (2007) Climate change (2007), fourth assessment report. Cambridge University Press, Cambridge

Jaetzold R, Schmidt H, Hornet ZB, Shisanya CA (2007) Farm management handbook of Kenya. In: Natural conditions and farm information, 2nd edn, vol 11/C. Ministry of agriculture/GTZ, Nairobi

Jury MR (2002) Economic impacts of climate variability in South Africa and development of resource prediction models. J Appl Meteorol 41:46–55

Kimani SK, Nandwa SM, Mugendi DN, Obanyi SN, Ojiem J, Murwira HK, Bationo A (2003) Principles of integrated soil fertility management. In: Gichuri MP, Bationo A, Bekunda MA, Goma HC, Mafongoya PL, Mugendi DN, Murwuira HK, Nandwa SM, Nyathi P, Swift MJ (eds) Soil fertility management in Africa: a regional perspective. Academy Science Publishers (ASP), Centro Internacional de Agricultura Tropical (CIAT), Tropical Soil Biology and Fertility (TSBF), Nairobi, Kenya, pp 51–72

Kosgei JR (2008) Rainwater harvesting systems and their influences on field scale soil hydraulic properties, water fluxes and crop production. Ph.D thesis, University of KwaZulu-Natal, Pietermaritzburg, South Africa

Kumar KK, Rao TVR (2005) Dry and wet spells at Campina Grande-PB. Rev Brasil Meteorol 20 (1):71–74

Li L, Zhang S, Li X, Christie P, Yang S, Tang C (2003) Inter-specific facilitation of nutrient uptakes by intercropped maize and faba bean. Nutr Cycl Agro-ecosyst 65:61–71

MATLAB (Matrix Laboratory) (2013) The empirical cumulative distribution function (ECDF). Web: http://www.mathworks.com/help/stats/ecdf.html. Accessed 12 July 2013

Meehl GA, Stocker TF, Collins WD, Friedlingstein P, Gaye AT, Gregory JM, Kitoh A, Knutti R, Murphy JM, Noda A, Raper RL, Watterson IG, Zhao ZC (2007) Global climate projections. In: Solomon S, Qin D, Manning M, Chen Z, Marquis M, Averyt KB, Tignor M, Miller HL (eds) Climate change (2007): the physical science basis. Contribution of working group I to the 4th assessment report of the intergovernmental panel on climate change. Cambridge University Press, Cambridge

Micheni AN, Kihanda FM, Warren GP, Probert ME (2004) Testing the APSIM model with experiment data from the long term manure experiment at Machang'a (Embu), Kenya. In Delve RJ, Probert ME (eds) Modeling nutrient management in tropical cropping systems. Australian Center for International Agricultural Research (ACIAR) No 114, Canberra, pp 110–117

Mugwe J, Mugendi D, Odee D, Otieno J (2009) Evaluation of the potential of organic and inorganic fertilizers of the soil fertility of humic hit sol in the central highlands of Kenya. Soil Use Manag 25:434–440

Mzezewa J, Misi T, Ransburg L (2010) Characterization of rainfall at a semi-arid ecotope in the Limpopo province (South Africa) and its implications for sustainable crop production. Web: http://www.wrc.org.za. Accessed 7 June 2013

Nicholson SE (2001) Climatic and environmental change in Africa during the last two centuries. Clim Res 17:123–144

Raes D, Willems P, Baguidi F (2006) RAINBOW: a software package for analyzing data and testing the homogeneity of historical data sets. In: Proceedings of the 4th international workshop on 'sustainable management of marginal dry lands'. Islamabad, Pakistan, 27–31 Jan 2006

Recha CW, Makokha GL, Traore PS, Shisanya C, Lodoun T, Sako A (2012) Determination of seasonal rainfall variability, onset and cessation in semi-arid Tharaka district, Kenya. Theoret Appl Climatol 108:479–494. doi: 10.1007/s00704-011-0544-3

Rockstrom J, Barron J, Fox P (2002) Rainwater management for increased productivity among small-holder farmers in drought prone environments. Phys Chem Earth 27(11–22):949–959

Seleshi Y, Zanke U (2004) Recent changes in rainfall and rainy days in Ethiopia. Int J Climatol 24:973–983

Shisanya CA (1990) The 1983–1984 drought in Kenya. J East Afr Resour Dev 20:127–148

Sivakumar MVK (1991) Empirical-analysis of dry spells for agricultural applications in SSA Africa. J Clim 5:532–539

Sivakumar MVK (1992) Empirical analysis of dry spells for agricultural applications in West Africa. J Clim 5:532–539

Tilahun K (2006) Analysis of rainfall climate and evapo-transpiration in arid and semi-arid regions of Ethiopia using data over the last half century. J Arid Environ 64:474–487

Chapter 6
Addressing the Potential Impacts of Climate Change and Variability on Agricultural Crops and Water Resources in Pennar River Basin of Andhra Pradesh

Sridhar Gummadi and K.P.C. Rao

Abstract The objective of the current study is to address the possible potential impacts of climate change and variability on agricultural crops and water resources in Pennar river basin, of Southern India. As part of the study Integrated Modelling Assessment (IMS) was developed by establishing functional links between hydrological model Soil Water Assessment Tool (SWAT), agricultural crop simulation model Environmental Policy Integrated Climate (EPIC) and regional climate model Providing REgional Climates for Impacts Studies (PRECIS). Database pertaining to climatic parameters, hydrological and agro-meteorological inputs to run integrated assessment systems are synthesized to run the model for study area. The model in general aim at major driver of this study is HadRM3 (Hadley Centre third generation regional climate model)—The Hadley Center Regional Climate Models resolution, which is 0.44° × 0.44° (approx. 50 km cell–size) on ground covering an average size of typical Indian districts/sub-basins. For regional levels the results are obtained by aggregating from the sub-basin/district level. The assessment will include the following components: (1) Baseline climatology, (2) Under global warning HadRM3 derived climate change scenarios, (3) Water Resources (Hydrological) analysis including irrigation water, and (4) agro-meteorological analysis including soil-water regime, plant growth and cropping pattern. Overall in Pennar region results revealed that the mean annual flows in the river system would increase by 8 % in A2 and 4 % in B2 whereas, increase in evapotranspiration losses were found to be about 10 % in A2 and 12 % in B2. Impacts on crop yields is the combined effect of increased surface temperatures, decreased rainfall and higher ambient atmospheric CO_2. Three rain-fed crops (Groundnut, Sorghum, Sunflower) show decreased yields under A2, whereas B2 seemed to be relatively better than A2. The decrease is significant for groundnut (−38 % for A2

S. Gummadi (✉) · K.P.C. Rao
International Crops Research Institute for the Semi-Arid Tropics (ICRISAT), P.O. Box 5689, Addis Ababa, Ethiopia
e-mail: S.Gummadi@cgiar.org

W. Leal Filho et al. (eds.), *Adapting African Agriculture to Climate Change*, Climate Change Management, DOI 10.1007/978-3-319-13000-2_6

and −20 % for B2), but compared to groundnut impact were less detrimental for other two rain-fed crops (Sorghum and Sunflower). Rice being an irrigated crop in the region showed decrease in yield by −15 and −7 % for A2 and B2 scenarios respectively. Negative simulated crop yields in the region are predominantly due to increased surface temperatures in the future climate change scenarios.

Keywords Climate change · SWAT · EPIC · PRECIS

Introduction

Potential impacts of climate change on agricultural crops and water resources has utmost importance in tropical countries like India due to continues crop failures and shortage of water for domestic, industrial and agricultural purpose. It is well understood from the recent past studies that agricultural crops are most vulnerable to changes in weather and climate (Slingo et al. 2005; Osborne et al. 2007; Challinor and Wheeler 2008; Schlenker and Roberts 2008).

Agriculture is the backbone of India's economy and is highly dependent on the spatial and temporal distribution of monsoon rainfall. Much of the country relies on tropical monsoons for approximately 80 % of the annual rainfall and most of this falls within 3–4 months (Mitra et al. 2002). For most parts of India, this major proportion falls during the summer (June–September) monsoon season. The temporal and spatial variations of the Indian summer monsoon have great relevance in the context of agriculture, industrial development, and planning and policy formulation. The agriculture sector is expected to be significantly affected by a reduction in crop water availability and an increase in the probability of extreme weather events resulting from the combined influence of elevated CO_2 concentrations and rise in surface temperatures (Chiotti and Johnston 1995).

Addressing the impacts of climate change on agricultural crops and water resources are often based on dynamic crop, hydrological and climate simulations models. Simulation models are computer based representations of physiological process responsible for plant growth and development, evapotranspiration and partitioning of photosynthetic output to produce economic yield (Crop models) (Boote et al. 1998; Williams 1995; Challinor et al. 2005), hydrological models are physical process based models represent complex surface runoff, subsurface flow, channel flow and evapotranspiration (Arnold et al. 1998; Harding et al. 2012), Climate models are based on well-established physical principles and have been demonstrated to reproduce observed features of recent and past climate changes (Houghton et al. 2001; Gnanadesikan et al. 2006). General Circulation Models (GCMs) and Regional Circulation Models (RCMs) are the most effective approach to explore the processes in the atmosphere, ocean and land surface. Global climate models provide the starting point for construction of the current and projected changes in future climate due to the increased anthropogenic emissions. The coupled atmosphere-ocean general circulation models (AOGCMs) have become the best available tools in addressing and understanding future climate change projections (Houghton et al. 2001). The basic climate models (GCMs) are coupled with atmosphere (A) and ocean and sea-ice (O). The complex equations are solved using a 3D grid over the Earth‘s surface. The 21st century climate models have 5 major components of Earth‘s system atmosphere, ocean, land surface, cryosphere (sea ice, snow) and biosphere.

This study focuses on the projected changes in future climate and its impacts on water resources and agricultural crops grown in Pennar watershed of Andhra Pradesh, India. Although studies on the impacts of climate change and variability on agricultural productivity have been conducted at global and national levels, only a few studies have focused on the integrated impacts of climate change on agricultural crops and water resources. This study aims to investigate the potential impacts of climate on available water resources and agricultural crops using hydrological model (SWAT), agricultural model (EPIC) and regional climate model (PRECIS):

1. Evaluating the ability of existing crop (EPIC) and hydrological (SWAT) models in the study area to simulate current climate variability
2. To study the response of mean changes in future climate on crop production and surface flow in Pennar river basin and
3. Developing strategic adaptation measures in response to the negative impacts of climate in the study area

Materials and Methods

Site Description

The study is conducted in Pennar basin in Andhra Pradesh state of India. Pennar Basin extends over an area of 55,213 km^2, which is nearly 1.7 % of total geographical area of the country. The basin lies in the states of Andhra Pradesh (48,276 km^2) and Karnataka (6,937 km^2). Pennar River rises from the Chenna Kesava hills of the Nandi ranges of Karnataka and flows for about 597 km before out falling into Bay of Bengal. The principal tributaries of the river are the Jayamangal, the Kunderu, the Sagileru, the Chitravati, the Papagni and the Cheyyeru. The important soil types found in the basin are red soils, black soil, sandy soil and mixed soil.

It is generally flat, having mostly slopes of less than 6.5 %. The basin is divided into 58 sub-basins covering four districts namely Kurnool, Ananthpur, Cuddapah and Chiitor comprises 160 Mandals (district sub-units). This study is conducted for all the Mandals of above mentioned four district and sub-basin as shown in the Fig. 6.1 Study area is located between, 77.10–80.15°E and 13.3–15.8°N.

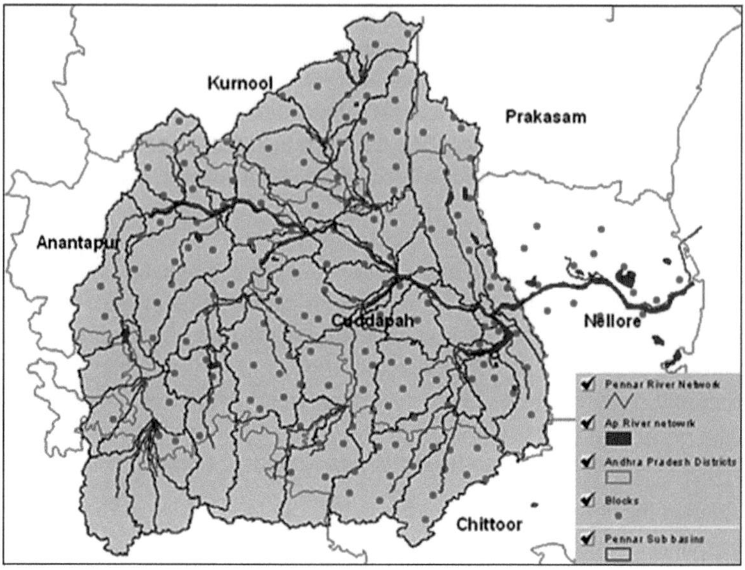

Fig. 6.1 Pennar basin—Andhra Pradesh, India

Digital Elevation Model (DEM)—DEM represents a topographic surface in terms of a set of elevation values measured at a finite number of points. Shuttle Radar Topography Mission (SRTM) ~90 m resolution DEM has been used for the study. Drainage relief, river network and rainfall stations of study area are shown in Fig. 6.2.

Climate is predominately semi-arid to arid. In general, there are four seasons in this region. Hot weather (from March to May), Southwest monsoon (from June to September), Northeast monsoon (from October to December) and winter (from

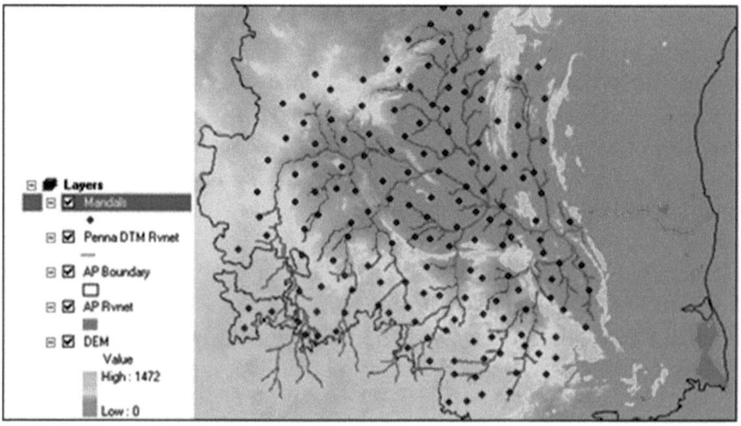

Fig. 6.2 Drainage and relief in Pennar basin, Andhra Pradesh

December to February). The state of Andhra Pradesh is divided into seven zones based on the agro-climatic conditions. The classification mainly concentrates on the range of rainfall received, type of the soils and topography. Study region falls in Rayalseema region including Ananthpur, Chittor, Cuddapah and Kurnool districts and parts of Prakasam and Karnataka state.

Climate data required by the models are daily precipitation, maximum/minimum air temperature, solar radiation, wind speed and relative humidity. These daily climatic inputs are entered from historical records in the model using monthly climate statistics that are based on long-term weather records. In this study, historical precipitation and temperature records for Pennar basin are obtained for 4 Indian Meteorological Department (IMD) weather stations located in and around the watershed are used for the current study. Stations Names are Kurnool, Anantapur, Cuddapah and Chittor in which the study area lies. Rayalaseema zone is in the semi-arid track. It receives an average annual rainfall in the range of 500–1,000 mm. Most of which come from southwest monsoon and the northeast monsoon. The rains normally begin in the second week of June and lasts till September (Southwest monsoon), which marks the main growing season (locally known as *Kharif*).

Results

Model Evaluation

Hydrological Model—SWAT

SWAT hydrological model is validated and calibrated over a 15-year period (1988–2003) by using historical climate data and comparing simulated output with the observed stream flows measured at four gauge stations in the basin. SWAT simulation methodology consist of an initial calibration and then followed by a second phase in which the impacts of climate change is to be assessed. The following model options are used for all the simulations performed (1) CN method for portioning of precipitation between surface runoff and infiltration, (2) Masking method for channel routing and (3) Penman Monteith method for potential evapotranspiration.

SWAT model runs are performed basically for two sets of rainfall data viz., (1) IMD rainfall and (2) Block rainfall data. IMD runs made use of the data for 4 stations where as block level data made use of 120 stations of rainfall data. Rainfall data for the period 1985–1995 has been used for IMD runs. Block level runs made use of the rainfall for the period 1988–2002. Other data sets viz., Soil, temperature and weather data remains same in both the runs. Flow data sets used for calibration include, Upper Pennar Reservoir—1971–2000, Tadipatri 1974–1998, Pennar Anicut—1983–1991, Somasila Reservoir—1979–1993. It was observed that flow

Table 6.1 Hydrological calibration parameters and calibrated values

Calibration parameter	Symbol	Initial estimate	Calibrated value
Curve no. of moisture cond. II	CN2	Estimated using AVSWAT	+6.6 %
Soil available water capacity	SOL_AWC	0.0	0.04

data is intermittent with long periods of no flow. Hence model calibration has been done for Annual and Monthly Runoff Comparisons for Tadipatri, Annual Runoff Comparison for Somasila and Daily flow comparison for Pennar Anicut. The parameters selected for calibration are shown in Table 6.1. Parameters were allowed to vary during calibration process within acceptable ranges across the basin until acceptable fit between the measured and simulated values are obtained at gauge locations.

SWAT simulated surface runoff for Tadipatri is compared with recorded runoff and it is noticed that the model simulated surface runoff is in good agreement with the observed runoff as depicted in Fig. 6.3.

Average annual rainfall is about 660 mm historically; it increased to 709 mm in A2 scenario and 683 mm in B2 scenario. There is an about 8 % increase in rainfall in A2 and about 4 % increase in rainfall in B2 scenario. It is observed that the runoff in the basin is varied from 4 to 11 %. Evapotranspiration losses are high. It varied from 80 to 95 %. In the climate change scenario, runoff in percentage of rainfall is about 19 % in A2 and 15 % in B2. In the climate change scenario, study estimated that the mean annual flow in the river system would be increased by 8 % in A2 and 4 % in B2. Evapotranspiration losses were decreased by about 10 % in A2 and 12 % in B2. The flows showed high inter-annual variability, which in turn reduce the river flow in dry years significantly, which would have serious effects on irrigation supply. An average rainfall increase of 4–8 % caused a 10–15 % increase in river flows. This may be due to an estimated wet condition in the climate change

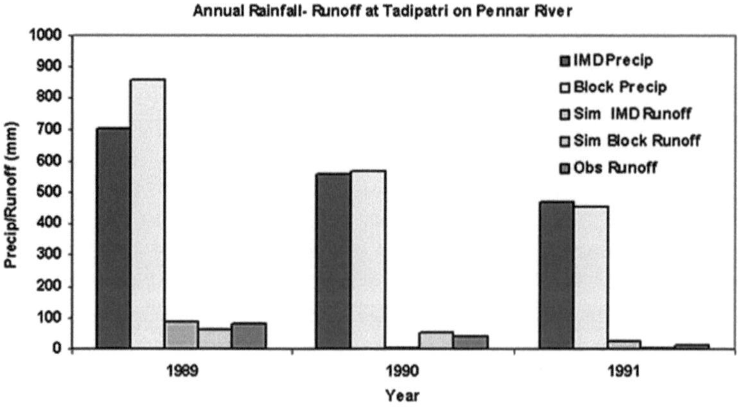

Fig. 6.3 Comparison of annual simulated and observed runoff at Tadipatri

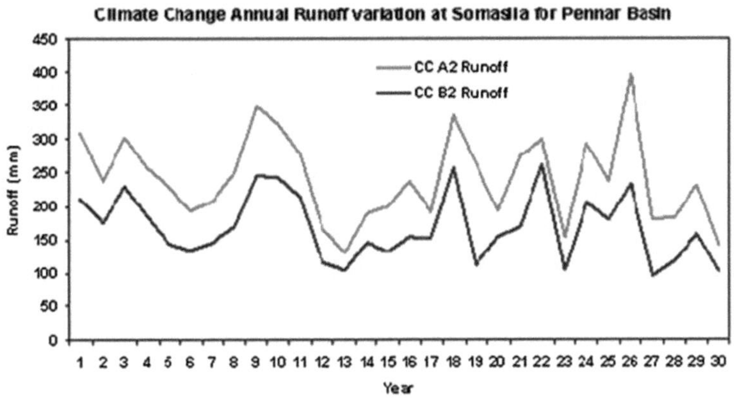

Fig. 6.4 Inter annual variability of runoff in future projected climate change scenarios

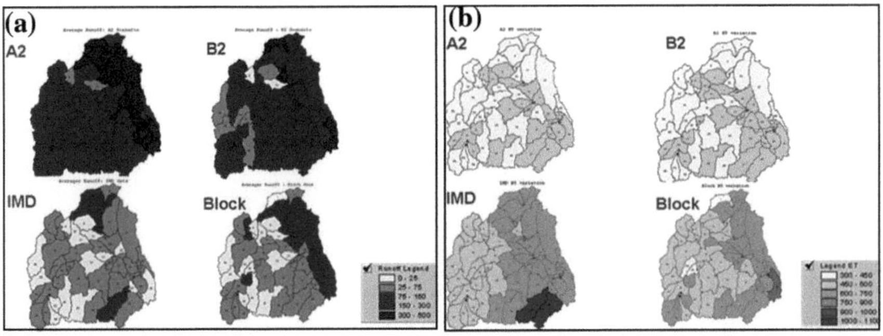

Fig. 6.5 Spatial distribution of **a** average annual runoff and **b** evapotranspiration in climate change scenarios of Pennar river basin

scenario. In A2 scenario, there is about 20 % chance that the rainfall exceeds by 1σ and 4 % exceeds by 2σ. Similarly number of instances in which rainfall is below 1σ is 14 % and 2σ is 4 %. The corresponding numbers in B2 scenario are 18, 6, 14 and 2 % respectively. These values indicate that the extremities in runoff will relatively high in A2 than B2 as shown in Fig. 6.4.

Spatial distribution of runoff and evapotranspiration across the basin in the climate change scenarios are shown in Fig. 6.5. These changes are not uniform across the basin. Increase in runoff is more significant in northern portion of the basin. This is the region, which showed relatively high rainfall and low evapotranspiration over other regions of the basin.

Agricultural Model—EPIC

Agriculture, like rest of India, is the main activity in the Pennar basin. The major food grains grown include rice, groundnut, sorghum, maize and sunflower.

Table 6.2 Crop calendar for Pennar river basin

Jun	Jul	Aug	Sep	Oct	Nov	Dec	Jan	Feb	Mar	Apr
Kharif						Rabi				
South-west monsoon				North-east monsoon		Winter			Summer	
Kharif (rainfed/irrigated rice)						Rabi (irrigated rice)				

Sugarcane, cotton and a variety of other pulses are also grown. These crops are grown either under irrigation or rainfed or both. The area is characterized by two growing season. The major crop growing season is between June and September (*Kharif*). The major source of water for crop production is the rainfall from southwest monsoon during June to September. The second growing season starts in December and last until April (*Rabi*). The main crop grown during *Kharif* season is rice and the source of water is irrigation. It is mostly pumped by electrical driven sub-mersible pumps from the ground source. Table 6.2 shows the relevant cropping calendar.

The four crops rice, groundnut, sunflower, and sorghum are selected for analysis in this study which are already been included in EPIC simulation model, but needed to be modified to reflect local conditions. The model was run for all four crops for *Kharif* season only. Except Rice remaining three crops are rainfed. Rice being an irrigated crop simulation is carried out based on the prevailing conditions in the field. About 47 parameters related to crop phenology, its environment and crop growth in a stressed environment are used in EPIC. Parameter values for the selected crops and the management practices associated with them are based on previous modeling exercises with EPIC and on advice from experts at the Acharya N. G. Ranga Agricultural University (ANGRAU) Hyderabad. EPIC simulated yields are generated at adminstrative blocks falling under four major districts (Kurnool, Chuddapah, Chittor and Ananthpur) of Pennar basin and database developed to describe agricultural practices and environmental conditions in each of these 160 blocks are being used. Soil properties are derived from the National Bureau of Soil Survey and Land use planning (NBSS&LUP) Nagpur paper maps at 1:250 K scale are employed. Validation of crop simulation model EPIC is carried out at districts level. EPIC is forced at block level and yields are aggregated to district level for the years 1989 through 1996 and the annual reported yields for the selected four crops viz., rice, sorghum, groundnut and sunflower. The validation was done using *Kharif* simulated crop yield, which were compared with annual (Kharif + Rabi) reported yields, which were the only data available. The crops, other than rice, are majorly a dryland crop dependent on southwest monsoon, extent of irrigation crops under Rabi season have not been covered in this study. Nevertheless, the validation test is still powerful since a predominance of annual yield is derived from the Kharif season. For instance statistical analysis on crop growing region shows that in the Ananthpur district of Andhra Pradesh the area planted in the Kharif versus rabi season were for rice 2.7 times, and groundnut 41 times. Rice tended to be irrigated in both seasons.

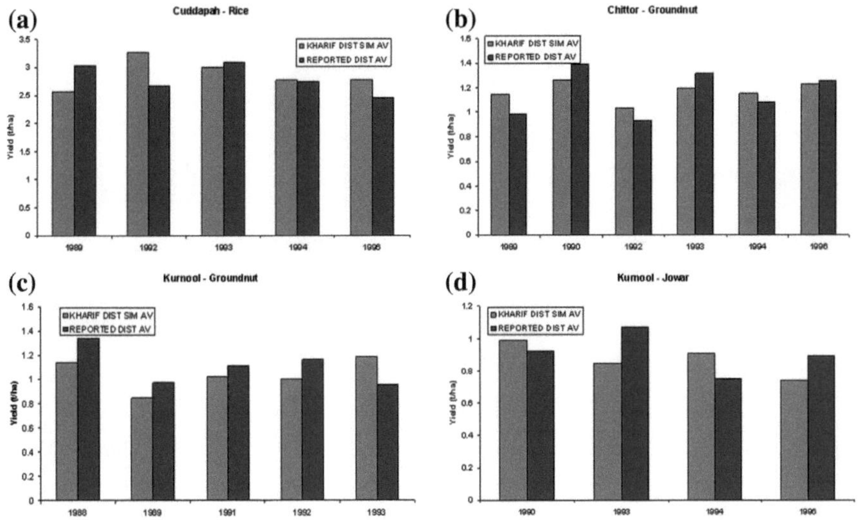

Fig. 6.6 Validation of epic crop simulation model for **a** rice crop at Cuddapah district, **b** groundnut at Chittor district, **c** groundnut at Kurnool district and **d** Sorghum at Kurnool district of Andhra Pradesh

Few examples of closeness between reported and simulated yield can be seen in Fig. 6.6 through 4.7, while performing all these simulations various intermediate checks have been performed, which helped achieve estimated yield close to reported yield.

Addressing the potential impacts of climate change on the four agricultural crops in the Pennar river basin, PRECIS simulated climate change scenarios are downscaled using Delta method. In this study a delta downscaling method is carefully chosen for its proven robust and popular, most likely because it is straightforward and relatively easy to understand. Delta method calculates changes in surface temperatures (ΔT) and relative changes in precipitation (ΔP) and perturb the projected changes to observed climate data and B2 scenarios respectively.

Table 6.3 shows the climate change scenarios for the Pennar region developed by perturbing the projected changes to historical climate data at Block level for both

Table 6.3 Precis projected climate change scenarios for Pennar region

Scenarios	Period	Changes in max temp (°C)			Changes in min temp (°C)			% Changes in RF (mm)		
		Highest	Lowest	Mean	Highest	Lowest	Mean	Highest	Lowest	Mean
A2	Kharif	3.5	3.0	3.1	3.4	3.1	3.2	20.8	−4.5	8.1
B2	Kharif	2.5	2.1	2.3	2.6	2.3	2.4	3.9	−12.0	-5.7
A2	Annual	3.3	2.9	3.1	3.7	3.4	3.6	28.2	9.8	21.3
B2	Annual	2.3	2.0	2.2	2.7	2.5	2.6	7.7	1.0	4.1

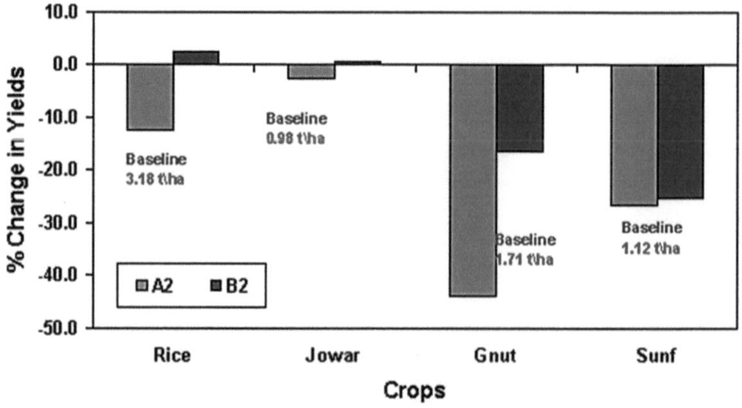

Fig. 6.7 Projected future changes in crop yields over Pennar river basin in a climate change scenario

A2 and B2 scenarios. The A2 scenario shows 21 % increase in the annual mean rainfall in the region and B2 scenario it is about 4 %. Increase in the seasonal mean rainfall for A2 scenario is about 8 % with −5 to 21 % variation, whereas it is −6 % with a range of −12 to 4 % for B2. The region will experience about 3 °C raise in the annual maximum temperature in A2 and 2 °C in B2, respectively. The warming trend will be in the range of 2.9–3.3 °C in A2 and 2.0–2.3 °C in B2. In case of minimum temperature about 3.6 °C raise in A2 and 2.6 °C in B2, respectively. The annual minimum temperature range would be between 3.4 and 3.7 °C in A2 and 2.5 and 2.7 °C in B2.

Under the regional climate change scenarios (both A2 and B2), groundnut showed highest negative deviation, where decrease in the yield appears to be −40 and −19 % for A2 and B2 scenarios respectively.

Following this sunflower showed nearly −18 and −16 % reduction in yield for A2 and B2 scenarios, sorghum varied between −5 and +1 % as shown in Fig. 6.7. Rice seems to have less impact with −11 % reduction decrease in yield under A2 while B2 seemed to have marginal positive impact with +2 % increase in the yield. Change in yield vary within the region due to changes in climate and other key inputs like crop management, soil and topography.

Conclusions

Agriculture represents a core part of the Indian economy and provides food and livelihood activities to a major portion of the Indian population. While the magnitude of the impacts of climate change varies as per region, climate change generally has an impact on agricultural productivity and shifting crop patterns.

Unfortunately, crop agriculture is highly dependent on degrading land quality and most importantly, dwindling and precarious water resource availability. The hydrological features of the region, especially in the Indian sub-Continent, are influenced by monsoons, and to a certain extent, on Himalayan Glacial melt. Under climate change scenario, the spatial-temporal behavior of monsoon will change significantly.

Results revealed that the mean annual flows in the river system would increase by 8 % in A2 and 4 % in B2 whereas increase in evapotranspiration losses were found to be about 10 % in A2 and 12 % in B2. Impact on yields is the combined effect of increase temperature, decreased rainfall and increased CO_2. Three rain-fed crops (Groundnut, Sorghum, Sunflower) show decreased yields under A2, whereas B2 seemed to be relatively better than A2. The decrease is significant for groundnut (−38 % for A2 and −20 % for B2), but compared to groundnut impact were less detrimental for other two rain-fed crops (as shown in graph below). Rice being an irrigated crop shows a decrease in yield by −15 and −7 % for A2 and B2 scenarios respectively. Decrease in yields are mainly due to the further increase in temperature under CC scenarios, as has also been observed in closed and open field experiments.

References

Arnold JG, Srinivasan A, Muttiah RS, Williams JR (1998) Large area hydrologic modeling and assessment part I: model development. J Am Water Resour Assoc 34(1):73–89

Boote KJ, Jones JW, Hoogenboom G (1998) Simulation of crop growth: CROPGRO model. In: Peart RM, Curry RB (eds) Agricultural systems modelling and simulation. M. Dekker, New York, pp 651–691

Challinor AJ, Slingo JM, Wheeler TR, Doblas-Reyes FJ (2005) Probabilistic hind casts of crop yield over western India. Tellus 57A:498–512

Challinor AJ, Wheeler TR (2008) Crop yield reduction in the tropics under climate change: processes and uncertainties. Agric For Meteorol 148:343–356

Chiotti QP, Johnston T (1995) Extending the boundaries of climate change research: a discussion on agriculture. J Rural Stud 11:335–350

Gnanadesikan A et al (2006) GFDL's CM2 global coupled climate models. Part II: the baseline ocean simulation. J Clim 19:675–697

Harding BL, Wood AW, Prairie JR (2012) The implications of climate change scenario selection for future stream flow projection in the upper colorado river basin. Hydrol Earth Syst Sci Discuss 16:3989–4007. doi:10.5194/hess-16-3989-2012

Houghton JT, DingY, Griggs DJ, Noguer M, Van der Linden PJ, Dai X, Maskell K, Johnson CA (2001) Climate change 2001. The scientific basis. contribution of working group I to the third assessment report of the intergovernmental panel on climate change. Cambridge University Press, Cambridge, p 881

Mitra AP, Kumar D, Rupa M, Kumar K, Abrol YP, Kalra N, Velayutham M, Naqvi SWA (2002) Global change and biogeochemical cycles: the South Asia region. In: Tyson P, Fuchs R, Fu C, Lebel L, Mitra AP, Odada E, Perry J, Steffen W, Virji H (eds) Global-regional linkages in the earth system. Springer, Berlin

Osborne TM, Lawrence DM, Challinor AJ, Slingo JM, Wheeler TR (2007) Development and assessment of acoupled crop–climate model. Glob Change Biol 13:169–183

Schlenker W, Roberts MJ (2008) Nonlinear temperature effects indicate severe damages to U.S. crop yields under climate change. PANS 106:15594–15598
Slingo JM, Challinor AJ, Hiskins BJ, Wheeler TR (2005) Introduction: food crops in a changing climate. Philos Trans R Soc B 360:1983–1989
Williams JR (1995) The EPIC model. In: Singh VP (ed) Computer models of watershed hydrology. Water Resources Publisher, USA, pp 909–1000

Chapter 7
Grain Yield Responses of Selected Crop Varieties at Two Pairs of Temperature Analogue Sites in Sub-humid and Semi-arid Areas of Zimbabwe

Justice Nyamangara, Esther N. Masvaya, Ronald D. Tirivavi and Adelaide Munodawafa

Abstract Climate analogues, based on 30 years meteorological data, were identified in smallholder areas of Zimbabwe. The sites were Kadoma (722 mm annual mean rainfall; 21.8 °C annual mean temperature) which was the higher temperature analogue site for Mazowe (842 mm annual mean rainfall; 18.2 °C annual mean temperature) for wetter areas, and Chiredzi (541 mm annual mean rainfall; 21.3 °C annual mean temperature) which was the higher temperature analogue site for Matobo (567 mm annual mean rainfall: 18.4 °C annual mean temperature) for drier areas. At each site and for each crop, three varieties were laid out in a randomized complete block design with three replications. The trials were conducted for two seasons (2011/2012 and 2012/2013). Maize and groundnut yields were higher at the cooler and wet sites and decreased significantly at the warmer and dry sites. In case of sorghum and cowpea, yields at the hotter site remained high implying that these crops are more tolerant to warmer temperatures predicted for 2050. At the drier sites, yields for all crops were significantly lower at the hotter site implying that crop production in the 2050s climate of the cooler site will be more difficult. The hypothesis that with increasing surface temperatures in a climate change scenario short duration genotypes can perform better compared with long duration was not confirmed.

Keywords Climate change · Crop varieties · Food security · Rainfall pattern · Temperature

---

J. Nyamangara (✉) · E.N. Masvaya · R.D. Tirivavi
International Crops Research Institute for the Semi-Arid Tropics (ICRISAT),
Matopos Research Station, P.O. Box 776, Bulawayo, Zimbabwe
e-mail: j.nyamangara@cgiar.org

A. Munodawafa
Department of Land and Water Resources Management, Faculty of Natural Resources Management and Agriculture, Midlands State University, Private Bag 9055, Gweru, Zimbabwe

W. Leal Filho et al. (eds.), *Adapting African Agriculture to Climate Change*,
Climate Change Management, DOI 10.1007/978-3-319-13000-2_7

Introduction

Crop productivity and food systems are predicted to be affected by changing climate which is likely to affect crop variety preferences by farmers across varying agro-ecological regions in future (Gregory et al. 2005). In Zimbabwe, conditions for growing early maturing and relatively lower yielding maize varieties are projected to shift more into currently wetter regions experiencing changing conditions suitable for growing long duration and relatively higher yielding varieties (Nyabako and Manzungu 2012). Reduction of crop yields is likely to result in a fall in crop revenue by as much as 90 % (Carter et al. 2007). The changes in crop production patterns are more likely to affect the marginalized smallholder farmers, who already experience low productivity due to current socio-economic and biophysical challenges characterizing the drier areas of sub-Saharan Africa (SSA) thereby impacting negatively on food security (Matarira et al. 1995). These predicted changes call for a focus on adaptive cropping strategies that will serve as mitigation measures against drastic changes in smallholder farmers' livelihoods (Eriksen et al. 2011).

Exposing smallholder farming communities to various crop variety options than those they traditionally grow might be a way forward in preparing them for the future. Days to maturity vary from crop to crop and are influenced by crop genotype, climatic and environmental factors (Bruns 2009). As the conditions in the wetter areas get drier and warmer, it is important for farmers to realize that they can no longer continue to grow high yielding crop varieties that take long to mature but rather should move to shorter duration varieties to ensure food security. Although crop yields are predicted to fall by adoption of shorter season varieties, the quantum is expected to be lower compared with that from continuing to grow longer season varieties which can completely fail to mature if the rainy season ends prematurely as predicted to happen more frequently in much of SSA in future.

It is against this background that trials to assess the grain yield of selected varieties of four crops were established at two pairs of analogue sites (wetter and drier) differing in temperature (2–4 °C) but with similar rainfall patterns. The crops were selected based on farmer preferences and suitability to different climatic conditions whilst the varieties were selected based on the number of days to maturity. It was hypothesized that as temperature increases in climate change scenario, shorter duration crop varieties will substitute longer duration varieties.

Materials and Methods

Site Description

The trials were conducted at two climate analogue sites. The wetter pair was Mazowe (cool/wet, 842 mm annual rainfall; 18.2 °C annual mean temperature) and Kadoma (hot/wet, 722 mm annual rainfall; 21.8 °C annual mean temperature), and

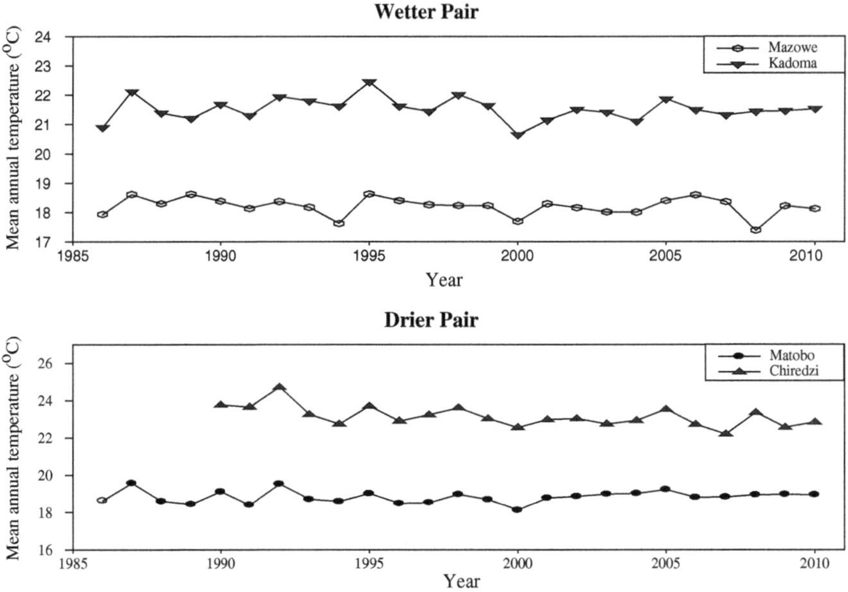

Fig. 7.1 Long-term temperature variability at the wetter and drier analogue pairs used for the study

the drier pair was Matobo (cool/dry, 567 mm annual rainfall: 18.4 °C annual mean temperature) and Chiredzi (hot/dry, 541 mm annual rainfall; 21.3 °C annual mean temperature). Inter-annual mean temperature variability at the four sites is presented in Fig. 7.1. The soils at Mazowe and Kadoma sites are red clays derived from dolerite and are relatively more weathered and leached. The soils at the Matobo site are loamy sands derived from granite rocks and those from Chiredzi are sandy loams derived from a mixture of siliceous gneiss and mafic rocks.

Experimental Design, Management and Data Collection

A completely randomized block design is employed with each of the crop varieties replicated three times. The trials were implemented for two seasons (2011/2012 and 2012/2013). The crop varieties used in the trials are given in Table 7.1. Soil samples (0–0.15 m) were collected at the beginning of the first season (2011/2012) from three sites for laboratory testing and the results are presented in Table 7.2. The experimental sites were tilled using a tractor drawn plough before planting and red clay soils at Mazowe and Kadoma were also disked to break large clods. Maize was planted at a spacing of 0.9 m × 0.3 m (37,037 plants ha^{-1}), sorghum 0.75 m × 0.2 m (66,667 plants ha^{-1}), and groundnut and cowpea 0.45 m × 0.15 m (148,148 plants ha^{-1}). Plot sizes were 6.5 m by 6.5 m and yield was determined from 4 m by 4 m

Table 7.1 Crop varieties

Crop	Varieties		
	Early maturing	Medium maturing	Late maturing
Maize	SC403	SC513	SC727
Sorghum	Macia	SDSL89473	Pato
Groundnut	Nyanda	Natal common	Makhulu red
Cowpea	CBC1	CBC2	Landrace

Table 7.2 Soil characterization on analogue and reference sites

Site	pH (1 M $CaCl_2$)	Olsen-P (mg kg^{-1})	Total P (%)	Mineral N (mg kg^{-1})	Total N (%)	Organic C (%)
Matobo	5.3	0.1	0.01	3.7	0.04	0.8
Mazowe	5.6	0.5	0.1	2.4	0.1	1.6
Kadoma	6.1	0.5	0.04	3.7	0.1	1.3

net plots. Maize and sorghum were fertilized with 300 and 286 kg ha^{-1} basal fertilizer (7 %N:6 %P:6 %K) for the wetter and drier sites respectively and a top dressing fertilizer rate of 150 kg ha^{-1} (34.5 %N) was applied at 4–6 weeks after planting depending on rainfall pattern. Groundnut and cowpea received similar basal rates to maize and sorghum but were not top-dressed with N fertilizer; gypsum was applied to groundnut at flowering at 250 kg ha^{-1}.

All the plots were weeded three times each season using hand-hoes with the first weeding performed two weeks after planting. Armyworm (*Spodoptera exempta*) and other leaf eaters were controlled by spraying carbryl (1-naphthyl methylcarbamate) 85 % WP. Aphids were controlled in cowpea by spraying diamethoate (O, O-Dimethyl S-(N methylcarbamoylmethyl) phosphorodithioate). Harvesting was done at physiological maturity and yields were adjusted to moisture content by 10 and 12.5 % for legumes and cereals, respectively.

Statistical Analysis

Grain yields were analyzed using analysis of variance (ANOVA) in GenStat 14th edition (2011). The standard error of differences (SED) of the means ($P < 0.05$) was used to separate site and variety means.

Results

Figure 7.2 shows that in the first season (2011/2012) Matobo received the least amount of rainfall which was 66.5 % lower than the hotter analogue pair (Chiredzi). In the second season Kadoma received 74.7 % lower rainfall than the cooler

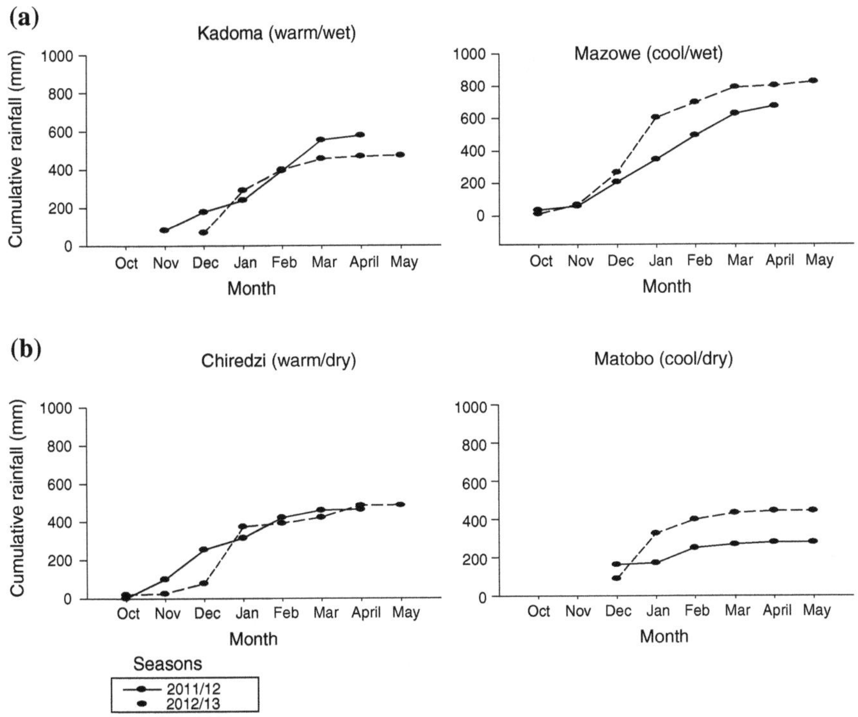

Fig. 7.2 Cumulative monthly rainfall recorded at the wetter and drier analogue sites during 2011/2012 and 2012/2013 cropping seasons. **a** Warm/wet and cool/wet analogue sites. **b** Warm/dry and cool/dry analogue sites

analogue pair (Mazowe) while the difference between the drier analogue pair was only 9.9 %. The soils from Mazowe and Kadoma were relatively more fertile as shown by higher pH, soil organic carbon and total nitrogen compared with Matobo (Table 7.2). Soil samples were not collected at the start of the first cropping season in Chiredzi.

Grain Yields at the Wetter Analogue Pair

In the first season maize yields followed the order long>short>medium season but the varietal differences were not significant (Table 7.3). In both seasons, maize yields were significantly higher ($P < 0.001$) at the cooler site (Tables 7.3 and 7.4). However, in the second season, varietal differences and variety by site interaction were significant ($P < 0.01$). The medium season variety (SC513) achieved the lowest yields in both seasons. The results implied a reduction in maize yield as

Table 7.3 Maize and sorghum grain yields at Kadoma (hot/wet) and Mazowe (cool/wet), Zimbabwe in the 2011/2012 season

	Maize yield (kg ha^{-1})			Sorghum yield (kg ha^{-1})		
	SC403	SC513	SC727	Macia	SDSL89473	Pato
Kadoma	3,373	2,529	3,979	3,121	5,468	3,928
Mazowe	5,083	4,857	5,937	5,223	2,007	3,861
	P value		SED	*P* value	SED	
Site	0.005		547.1	0.088	251.7	
Variety	0.198		672.0	0.396	308.2	
Interaction	0.902		952.3	<0.001	435.9	

temperature increases, especially in the first season when rainfall amount between the cooler and hotter sites was comparable (Fig. 7.2).

In the first season sorghum yields were varying between sites and varieties but the interaction between site and variety was significant ($P < 0.001$) (Table 7.3). The short season variety achieved the highest yield at the cooler site while the medium season variety achieved the highest yield at the hotter site (Table 7.3), and the trend was reversed between the sites in the second season (Table 7.4). In the second season, in addition to varietal differences site and variety interaction was also significant. The results imply that sorghum yield was less affected by higher temperature, even in a low rainfall season, and therefore was more resilient and more suitable (than maize) to predicted higher temperature conditions by 2050s.

Groundnut yields were significantly different both across sites ($P < 0.001$) and varieties ($P < 0.05$), Higher yields were recorded at the cooler site in both seasons (Tables 7.5 and 7.6). The yields were much low at the hotter site due to late start and premature end of both seasons which affected pod filling. In the first season, cowpea was affected by a fungal disease at the cool site (Mazowe) hence no grain yields were obtained. In the second season, both site and variety effects, and their interaction, were significant with the highest yields being recorded at the hotter site for all three cowpea varieties (Table 7.6). The results imply that groundnut, a crop

Table 7.4 Maize and sorghum grain yields at Kadoma (hot/wet) and Mazowe (cool/wet), Zimbabwe in the 2012/2013 season

	Maize yield (kg ha^{-1})			Sorghum yield (kg ha^{-1})		
	SC403	SC513	SC727	Macia	SDSL89473	Pato
Kadoma	2,202	1,848	1,552	3,164	2,085	1,595
Mazowe	5,222	4,358	5,531	1,934	3,172	1,720
	P value		SED	*P* value	SED	
Site	<0.001		106.0	0.865	36.2	
Variety	0.002		129.8	<0.001	44.3	
Interaction	<0.001		183.6	<0.001	62.7	

Table 7.5 Groundnut and cowpea grain yields at Kadoma (hot/wet) and Mazowe (cool/wet), Zimbabwe in the 2011/2012 season

	Groundnut yield (kg ha^{-1})			Cowpea yield (kg ha^{-1})		
	Nyanda	Natal common	Makhulu red	CBC1	CBC2	Landrace
Kadoma	158	355	140	1,250	1,546	976
Mazowe	3,809	1,420	2,094	–	–	–
		P values	SED	*P* value		SED
Site		<0.001	259	–		–
Variety		0.015	317.2	–		–
Interaction		0.007	448.6	–		–

Table 7.6 Groundnut and cowpea grain yields at Kadoma (hot/wet) and Mazowe (cool/wet), Zimbabwe in the 2012/2013 season

Groundnut yield (kg ha^{-1})				Cowpea yield (kg ha^{-1})		
	Nyanda	Natal common	Makhulu red	CBC1	CBC2	Landrace
Kadoma	163	470	1,326	636	783	867
Mazowe	662	603	1,027	306	692	514
		P value	SED	*P* value		SED
Site		<0.001	36.4	0.018		94.1
Variety		0.007	44.5	0.086		115.3
Interaction < 0.001			63.0	0.476		163.0

which requires a relatively long season, will become less favorable in future as the temperature increases and rainfall reliability decreases. However cowpea will become more suitable as cooler wet areas become hotter.

Grain Yields at the Drier Analogue Pair

At the drier sites, maize yields were significantly higher ($P < 0.001$) at the cooler site (Matobo) in both seasons (Tables 7.7 and 7.8) although higher rainfall was recorded at the hotter site (Fig. 7.2). Variety and site by variety interaction effects

Table 7.7 Maize and sorghum grain yields at Chiredzi (hot/dry) and Matobo (cool/dry), Zimbabwe in the 2011/2012 season

	Maize yield (kg ha^{-1})			Sorghum yield (kg ha^{-1})		
	SC403	SC513	SC727	Macia	SDSL89473	Pato
Chiredzi	75	150	245	26	185	15
Matobo	2,523	2,152	3,450	2,374	2,337	1,781
		P values	SED	*P* values		SED
Site		<0.001	445.4	<0.001		173.5
Variety		0.435	545.6	0.235		212.5
Interaction		0.556	771.5	0.412		300.5

Table 7.8 Maize and sorghum grain yields at Chiredzi (hot/dry) and Matobo (cool/dry), Zimbabwe in the 2012/2013 season

Maize yield (kg ha^{-1})				Sorghum yield (kg ha^{-1})		
	SC403	SC513	SC727	Macia	SDSL89473	Pato
Chiredzi	469	270	266	1,133	967	348
Matobo	3,648	2,844	2,310	3,131	1,662	1,543

	P value	SED	*P* value	SED
Site	<0.001	109.4	0.018	473.9
Variety	<0.001	134.0	0.155	580.4
Interaction	0.004	189.5	0.544	820.7

were only significant in the second season. Sorghum yields were also significantly ($P < 0.05$) higher at the cooler site, and variety and site by variety interaction effects were not significant (Tables 7.7 and 7.8).

Groundnut yields were higher at the cooler site in both seasons but the differences were only significant in the second season (Tables 7.9 and 7.10). Variety and site by variety interaction effects were not significant in both seasons (Tables 7.9 and 7.10). Cowpea yields were significantly ($P < 0.001$) higher at the cooler site except for the long season variety which was higher at the hotter site (Table 7.10). Variety and site by variety interaction effects were only significant in the second season.

Table 7.9 Groundnut and cowpea grain yields at Chiredzi (hot/dry) and Matopos (cool/dry), Zimbabwe in the 2011/2012 season

	Groundnut yield (kg ha^{-1})			Cowpea yield (kg ha^{-1})		
	Nyanda	Natal common	Makhulu red	CBC1	CBC2	Landrace
Chiredzi	170	431	298	255	94	27
Matobo	410	432	416	857	1,167	1,153

	P value	SED	*P* value	SED
Site	0.303	110.2	<0.001	156.0
Variety	0.593	135.0	0.927	191.1
Interaction	0.685	190.9	0.358	270.2

Table 7.10 Groundnut and cowpea grain yields at Chiredzi (hot/dry) and Matopos (cool/dry), Zimbabwe in the 2012/2013 season

Groundnut yield (kg ha^{-1})				Cowpea yield (kg ha^{-1})		
	Nyanda	Natal common	Makhulu red	CBC1	CBC2	Landrace
Chiredzi	344	332	272	640	639	653
Matobo	520	467	385	982	1,032	380

	P value	SED	*P* value	SED
Site	0.009	45.4	0.001	36.5
Variety	0.207	55.6	<0.001	44.7
Interaction	0.853	78.7	<0.001	63.1

Discussion

At the wetter analogue sites, higher maize and groundnut yields were recorded at the cooler site. The higher grain yields at cooler site can be attributed to higher effective rainfall due to lower evapo-transpiration, especially in the first season when rainfall amount and distribution between the sites were similar. The lower maize and groundnut yields at the hotter site (Kadoma) imply that these crops will have a reduced contribution to food security in the Mazowe (cooler) in 2050s climate which is predicted to be hotter. Sorghum and cowpea yields at the hotter site were relatively high despite the higher temperature, even in a low rainfall season, and therefore will be more suitable (than maize and groundnut) in predicted 2050s climate in Mazowe. Currently sorghum is mostly grown in the drier parts of Zimbabwe (Rao and Mushonga 1987). Therefore there is need to prepare farmers in Mazowe, and similar agro-ecological areas, to adapt their future cropping systems to more drought tolerant crops such as sorghum and cowpea in order to reduce the impacts of climate change on their food security and income status. This will involve ensuring the availability of suitable crops and varieties, relevant technical advice, policy and institutional arrangements to ensure a sustainable value chain.

At the dry sites yields for all crops were higher at the cooler site (Matobo) compared with the hotter site (Chiredzi). In the second season yields for sorghum (967–1133 kg ha^{-1} for short and medium season varieties) and cowpea (639–653 kg ha^{-1}) at the hotter site were reasonable when compared to the national average (<500 kg ha^{-1} for sorghum) but still lower than at the cooler site (Matobo) where rainfall was much lower (Tables 7.7 and 7.8). The results imply that under the predicted 2050s climate for Matobo crop production will become more risky and therefore threatening the already low food security status of smallholder farmers. Sorghum and cowpea appear to be more resilient but there is need to develop even more tolerant varieties and to also explore the introduction of new crops which can withstand the predicted higher temperatures. There is need to consider alternative livelihood options for areas with predicted 2050s climatic conditions similar to those currently prevailing in Chiredzi. Livestock farming, supported by supplementary fodder production to complement natural grazing is one such option. There is also need to explore other off-farm livelihood options.

In the absence of related field studies in SSA modelling provides an alternative comparison. The results of our study are in agreement with estimates based on Global Circulation Models (GCMs) and crop simulation models (e.g. Rosenzweig et al. 1993; Matarira et al. 1995). National cereal yields for the 2050s are estimated to decrease by between 2.5 % and 30 % with and without CO2 (Parry et al. 2004). Using the CERES-Maize model under the geophysical fluid dynamics laboratory (GFDL), and the canadian climate centre model (CCCM) which estimate 2–4 °C increases in Zimbabwe and precipitation changes ranging from ± 10 to 15 % on average during the October to April season (Unganai 1996; Matarira et al. 1995) projected maize yield declines particularly in dry-land farming in some Zimbabwean communities by 2050.

Although the analogue pair concept provides valuable insights into future crop production scenarios it has got considerable limitations. For example, although rainfall amount may be similar at two sites, spatial and temporal variations between the sites occur and these may affect crop yield. Current agro-ecological conditions at a particular site impose certain growth conditions which limit the extent to which yields can be directly compared with another site.

Conclusions

We conclude that as temperature increases due to climate change, more drought tolerant crops such as sorghum and cowpea will become important in ensuring food security of smallholder farmers at the studied sites. For the drier sites, even drought tolerant crops will be inadequate to ensure food security of the farmers prompting the need to explore other tolerant crops or varieties or pursue livestock production or off-farm activities. The hypothesis that short season varieties will perform better as temperature increases was not confirmed.

Acknowledgments We would like to acknowledge the Deutsche Gesellschaft fur Internationale Zusammenarbeit (GIZ) on behalf of the Federal Republic of Germany who sponsored the project "Adapting agriculture to climate change: Developing promising strategies using analogue locations in Eastern and Southern Africa" (CALESA—Climate Analogue Locations in Eastern and Southern Africa).

References

Bruns HA (2009) A survey of factors involved in crop maturity. Agron J 101:60–66

Carter TR, Jones R, Lu X, Bhadwal S, Conde C, Mearns L, O'neill B, Rounsevell M, Zurek M (2007) New assessment methods and the characterisation of future conditions. Cambridge University Press, Cambridge

Eriksen S, Aldunce P, Bahinipati CS, Martins RDA, Molefe JI, Nhemachena C, O'brien K, Olorunfemi F, Park J, Sygna L (2011) When not every response to climate change is a good one: identifying principles for sustainable adaptation. Clim Dev 3:7–20

Gregory PJ, Ingram JS, Brklacich M (2005) Climate change and food security. Philos Trans R Soc B Biol Sci 360:2139–2148

Matarira C, Makadho J, Mwamuka F (1995) Zimbabwe: climate change impacts on maize production and adaptive measures for the agricultural sector. Interim report on climate change country studies, US country studies program, Washington, DC

Nyabako T, Manzungu E (2012) An assessment of the adaptability to climate change of commercially available maize varieties in Zimbabwe. Environ Nat Resour Res 2:32

Parry ML et al (2004) Effects of climate change on global food production under SRES emissions and socio-economic scenarios. Glob Environ Change 14:53–67

Rao SA, Mushonga J (1987) A catalogue of passport and characterization data of sorghum, pearl milletand finger millet germplasm from Zimbabwe. A catalogue of passport and characterization data of sorghum, pearl milletand finger millet germplasm from Zimbabwe

Rosenzweig C, Parry ML, Fischer, Frohberg K (1993) Climate change and world food supply. Research report no. 3. Environmental change unit. University of Oxford. pp 1 24
Unganai LS (1996) Historic and future climatic change in Zimbabwe. Clim Res 6:137–145
VSN (2011) GenStat for Windows, 14th edn. VSN International, Hemel Hempstead

Chapter 8
Adapting Agriculture to Climate Change: An Evaluation of Yield Potential of Maize, Sorghum, Common Bean and Pigeon Pea Varieties in a Very Cool-Wet Region of Nyandarua County, Central Kenya

Joseph M. Miriti, Anthony O. Esilaba, Karuturi P.C. Rao, Joab W. Onyango, Stephen K. Kimani, Peterson M. Njeru and John K. Lekasi

Abstract Three experiments were conducted to evaluate the performance of maize, sorghum, common bean and pigeon pea varieties under different water management in a cool and wet region of Central Kenya, as a part of the studies at analogue sites. The first experiment evaluated the growth and performance of three varieties (early maturing: EM, medium maturing: MM and late maturing: LM) of maize (*Zea mays* L), sorghum (*Sorghum bicolor L.*), pigeon pea (*Cajanus cajan*) and common bean (*Phaseolus vulgaris L.*). The second experiment evaluated maize and sorghum response to water conservation and three fertiliser rates (0, 20 and 40 kg N/ha). The third experiment assessed the effect of water conservation measures on crop yields of common bean and pigeon pea grown under three plant densities (low, medium and high). Tied ridge tillage was used as the water conservation measure and disc plough as the control in the second and third experiments. Maize, sorghum, pigeon pea and common bean took more than 180, 245, 217 and 95 days respectively, to reach physiological maturity. The MM maize variety (DK8031), EM pigeon pea

J.M. Miriti (✉) · S.K. Kimani · P.M. Njeru · J.K. Lekasi
Kenya Agricultural Research Institute, Muguga South, P.O. Box 30148-00200
Nairobi, Kenya
e-mail: jmmiriti@yahoo.co.uk

A.O. Esilaba
Kenya Agricultural Research Institute, P.O. Box 57811-00200
Nairobi, Kenya

K.P.C. Rao
International Crops Research Institute for the Semi-Arid Tropics (ICRISAT),
P.O. Box 823-00621, Nairobi, Kenya

J.W. Onyango
National Agricultural Research Laboratories, P.O. Box 14733-00800, Nairobi, Kenya

W. Leal Filho et al. (eds.), *Adapting African Agriculture to Climate Change*,
Climate Change Management, DOI 10.1007/978-3-319-13000-2_8

variety (ICPL 84091) and LM common bean variety (GLP 24) yielded the greatest grain of 4,938, 881 and 620 kg/ha respectively, among the respective crop varieties. The sorghum varieties were attacked by fungal and rust diseases leading to yield losses in all seasons. Soil water conservation in general did not have a significant effect on crop yield though there were yield improvements. In the plant density trial, the medium plant densities of pigeon pea (33,333 pl/ha) and common bean (148,148 pl/ha) resulted in the greatest grain yields. The highest grain yield of maize (4,184 kg/ha) and sorghum (47 kg/ha) was obtained in plots with 20 kg/ha of nitrogen fertilizer. Based on the results of this study, pigeon pea and common bean can be introduced in the farming systems to improve crop diversity. The production of the tested sorghum varieties should be discouraged in this region because they are prone to fungal and rust diseases due to the cold and wet weather conditions.

Keywords Climate change · Maize · Sorghum · Piegeonpea · Bean · Moisture conservation · Fertility management

Introduction

Global projections suggest that the ongoing climate change could have substantial impact on agricultural production and thus threatening food security by the end of the 21st century (IPCC 2007; Naab et al. 2013). Changes in temporal and spatial temperature and rainfall patterns will have great impacts on rain fed agriculture because of its direct dependence on rainfall and temperature. The crop water cycle is dependent on rainfall and changes in rainfall pattern both in time and quantity will directly impact on the crop development (Tao et al. 2003). Erratic rainfall impacts negatively on crop yield and the episodic food insecurity in Kenya is mainly contributed by season to season variability in rainfall. Temperature affects the function of plant physiological processes (photosynthesis, respiration, leaf expansion rate) during crop growth and yield development (Hay and Porter 2006). Variations in temperature and rainfall due to climate change in any location may have negative or positive effects on crop growth and yields resulting in food insecurity. To avoid climate-induced food shortages, adaptation to impacts of climate change is fundamental (Lobell et al. 2008).

The climate-induced food shortages can be avoided through more efficient irrigation, soil and water management, improved crop varieties and better land cultivation practices (Naab et al. 2013). The capacity for farmers to adopt these management practices depends upon access to information on season conditions and improved technologies that make best use of the season (Mutunga and Hardee 2009). Such information is often lacking because it is site-specific in terms of soil and climate conditions. Currently in Kenya, there is an ongoing research aimed at developing promising strategies for adapting agriculture to climate change using analogue locations.

The main goal of the research project is to improve the ability of rainfed farmers to adapt to progressive climate change through crop, soil and water management innovations and appropriate crop genotype choices. In order to achieve the object of this study, three trials were implemented: (1) to assess the growth and performance of an early, medium and late maturing varieties of maize, sorghum, pigeon pea and common bean, (2) to determine the effect of water conservation and fertiliser application on productivity of maize and sorghum, and (3) to determine the effect of water conservation and plant population on productivity of common bean and pigeon pea. This study was implemented in five sites with different climatic conditions from warm dry to very cool climate. One of the study sites selected for this study is Oljororok which has very cool and wet climate regime. This site is one of the major agricultural production areas in Kenya but it has been experiencing some rainfall and temperature changes due to climate change recently. Farmers have reported increased temperatures and more erratic rainfall patterns than before. The hypothesis of the study was that sorghum and leguminous crops such as pigeon pea and the common bean which are currently not grown in the humid areas may be the future crops for integrating the farming systems due to the emerging climate change. This paper presents some of the results obtained at this study site which is in Nyandarua County, central Kenya.

Materials and Methods

Study Site

The study was conducted at the Oljororok KARI experimental station from April 2012 to February 2014. The Oljororok site lies on latitude 0.04S and 36.35E at an elevation of 2,400 m above sea level. Its agro-ecological zone is described as Upper Highland sub-zone 2–3 (Pyrethrum, wheat and barley Zone) with average annual rainfall and temperatures of 820–990 mm and 14.0 °C, respectively (Jaetzold et al. 2006). Though the general rainfall pattern in Kenya is a bimodal type where the ‘long rains crop season’ fall between March and May and the ‘short rains crop season’ between October and December, the Oljororok site is very cool and wet, receiving rainfall throughout the year. This study was conducted between March 2013 and February 2014. The mean monthly rainfall and temperatures at the site during the study period were 0.6–295 mm and 13.9–19.4 °C, respectively (Table 8.1). The soil is classified as Ando luvic phaeozem which is deep dark brown soil, well drained with a clay loam texture (Jaetzold et al. 2006).

Table 8.1 Weather conditions at Oljororok during the study period (from March 2013 to February 2014)

Month	2012				2013				2014			
	Rainfall (mm)	Tmax (°C)	Tmin (°C)	RH (%)	Rainfall (mm)	Tmax (°C)	Tmin (°C)	RH (%)	Rainfall (mm)	Tmax (°C)	Tmin (°C)	RH (%)
Jan	0.0	22.8	5.1	43.5	55.3	24.8	8.1	58.1	0.6	22.7	7.0	69.0
Feb	29.8	23.3	6.3	49.4	0.0	23.4	7.0	48.5	55.8	23.1	8.7	59.2
Mar	3.5	24.9	6.9	41.7	124.6	26.7	12.1	56.6				
Apr	266.9	22.8	10.1	66.9	295.1	22.0	10.1	58.1				
May	206.3	21.3	8.9	73.4	106.7	21.7	6.9	48.5				
June	105.9	20.9	7.9	73.4	98.5	21.0	8.0	56.6				
July	173.8	19.6	8.6	78.9	214.1	22.7	11.5	72.3				
Aug	142.3	20.2	7.6	73.4	229.4	23.0	10.0	77.4				
Sep	80.0	21.5	6.9	68.2	66.8	23.3	11.4	67.3				
Oct	120.7	21.6	8.6	66.4	91.0	23.2	10.5	60.8				
Nov	34.6	21.1	9.4	67.6	130.3	20.3	9.7	73.7				
Dec	77.2	20.8	9.1	66.4	159.5	22.5	12.4	69.4				

*T*max maximum temperature, *T*min minimum temperature, *RH* relative humidity

Field Experiments

Assessment of the Performance of Cereal and Legume Crop Varieties Under Very Cool Climate

The objective of this trial was to assess the growth and performance of three varieties (early maturing: EM, medium maturing: MM and late maturing: LM) of maize, sorghum, pigeon pea and common bean under the very cool climatic conditions in Nyandarua County. The experimental design was a completely randomized block design (RCBD) replicated three times. Basal phosphorus fertilizer (40 kg N/ha) and nitrogen fertilizer (20 kg N/ha) was applied at the time of planting in all plots. The maize and sorghum plots were later top dressed with 20 kg N/ha after thinning. The trial was planted on 19th April 2012 and on 30th April 2013, during the long rains season of 2012 and 2013 using the crop varieties and plant spacing indicated in Table 8.3.

Determining the Effect of Water Conservation and Fertiliser Application on Productivity of Maize and Sorghum

The objective of this trial was to assess maize and sorghum response to soil moisture and soil fertility. The treatments consisted of two soil water management practices (flat tillage and tied ridges) and three N fertiliser rates (0, 20 and 40 kg N/ha). Treatments were arranged in a split-split plot design with tillage as main plots, N rates as sub-plots and crops as sub-sub-plots. All treatments were replicated three times. The plot size was 6 × 5 m. Maize (DK8031 variety) and sorghum (Kari Mtama 1 variety) were planted at a spacing of 75 by 30 cm (44,444 plants/ha) and 75 by 20 cm (66,666 plants/ha) respectively, on 16th April and on 30th April 2013, during the long rains season of 2012 and 2013, respectively.

Determining the Effect of Water Conservation and Plant Population on Productivity of Common Bean and Pigeon Pea

This trial assessed the effect of water conservation measures on crop yields of common bean (KK8) and pigeon pea (ICEAP 00557) grown under different plant densities. The soil water management practices evaluated were flat tillage and tied ridges. The treatments were arranged in a split-split plot design with tillage as main plots, plant densities as sub-plots and crops as sub-sub-plots. Under each soil management practice, the performance of 3 plant densities (low, medium and high) of pigeon pea and common bean were compared (Table 8.2). The plot size was 6 × 5 m replicated three times. Phosphorus fertilizer was applied at 40 kg P/ha in all plots at sowing. Pigeon pea and common bean were planted at a row spacing of 100 and 45 cm respectively.

Table 8.2 Plant population used for different treatments

Crop	Plant population		
	Low (pl/ha)	Medium (pl/ha)	High (pl/ha)
Common bean	111,111	148,148	185,185
Pigeon pea	25,000	33,333	41,666

Table 8.3 Days to physiological maturity and average yields (kg/ha) for different crop varieties

Crop variety	Variety name	Plant spacing (cm)	Days to maturity	Grain	TDM
Maize					
EM	DH04	75 × 30	180	3,810	9,777
MM	DK8031	75 × 30	189	4,938	14,457
LM	WH403	75 × 30	189	4,754	13,586
Sorghum					
EM	Macia	75 × 20	-	634	13,735
MM	Kari Mtama 1	75 × 20	-	203	9,044
LM	Gadam	75 × 20	-	1,117	11,219
Bean					
EM	Kat Bean 9	45 × 15	95	478	1,063
MM	KK8	45 × 15	111	464	1,868
LM	GLP 24	45 × 15	110	620	1,939
Pigeon pea					
EM	ICPL 84091	50 × 10	217	881	-
MM	ICEAP 00557	100 × 30	287	762	-
LM	ICEAP 00040	150 × 50	277	565	-

TDM above ground total dry matter, *EM* early maturing variety, *MM* medium maturing variety, *LM* late maturing variety

Results

Performance of Crop Varieties

The performance of maize, sorghum, common bean and pigeon pea varieties are given in Table 8.3.

Maize The average days to physiological maturity and the yield for maize varieties during the study period (LR2012 and LR2013 crop seasons). The EM variety matured at around 180 days after sowing and 10 days earlier the varieties. The maize crop growth cycle extended into the short rains seasons which start in October. Grain yield was highest with MM variety (4,938 kg/ha) followed by LM variety (4,757 kg/ha) and least with EM variety (3,810 kg/ha). The LM maize variety also out-yielded EM and LM varieties dry matter by 4,680 and 871 kg/ha, respectively.

Sorghum The MM variety (Kari Mtama 1) had the least grain yield value 203 kg/ha which was more than 67 % significantly ($P \leq 0.05$) lower than the yields for the other varieties (Table 8.3). No grain yield was obtained during the LR2013 season due to severe fungal and rust disease attack. The LM variety gave the highest average dry matter yield (13,735 kg/ha) which was 51 and 22 % more than dry matter yield for MM and EM maturing varieties, respectively. It was difficult to get the actual days to maturity for sorghum due to the rust disease which occurred during the grain filling stage and caused the grain to rot. From field observation, however, physiological started at around 245 days after sowing and harvesting was always done in February of the succeeding year.

Common bean The days to maturity and the yield data for common bean varieties in Table 8.3 is for three crop seasons (LR2012, SR2012 and LR2013). On average, LM variety took fewer days (95) to reach physiological maturity while maturity period for the other two varieties were similar at around 110 days after sowing. Significant varietal yield differences occurred in LR2012 season but it was dry matter yield only that showed significant differences in SR2012/13 season. The highest average grain yield were recorded with EM (620 kg/ha) followed by MM (464 kg/ha) and then LM (478 kg/ha) crop variety. Dry matter yield also followed grain yield trend. During the three seasons, the highest bean yield production occurred during the SR2012 and least in LR2012 season.

Pigeon pea The MM variety took the longest period (287DAS) to reach physiological maturity followed by LM and the EM variety in a decreasing order. Though the grain yields were not significantly different, the EM variety yielded the greatest grain (881 kg/ha) while the LM variety gave the least (565 kg/ha). We did not collect the dry matter yield because the harvesting of pods continued into the succeeding season and there was too much leaf fall. The EM variety was harvested in two flushes while the other varieties were harvested in one flush.

Effect of Water Conservation and Nitrogen Fertiliser Application on Crop Yield

Maize Effect of tillage and nitrogen application rates on maize yields is given in Figs. 8.1 and 8.2, respectively. Both tillage and N fertilizer did not have a significant effect on maize yield. Average maize grain yield under tied ridges was 3,977 kg/ha compared to 4,158 kg/ha under flat tillage system. Biomass yield under tied ridges and flat tillage were very almost equal at 10,955 and 10,802 kg/ha, respectively. When averaged across the tillage systems, the maximum grain (4,184 kg/ha) in plots with 20 kg/ha of N fertilizer and was 14 and 336 kg/ha more than the yields in plots with 0 and 40 KgN/ha, respectively. The greatest dry matter (10,996 kg/ha) was also obtained in plots with 20 KgN/ha fertilizer.

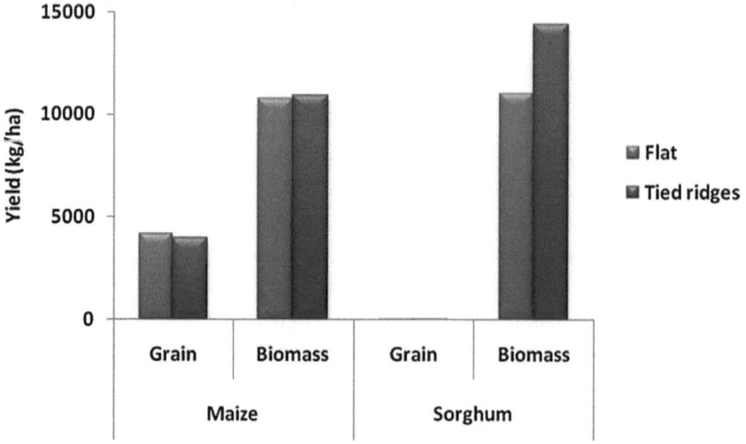

Fig. 8.1 Effect of water conservation tillage on maize and sorghum yields

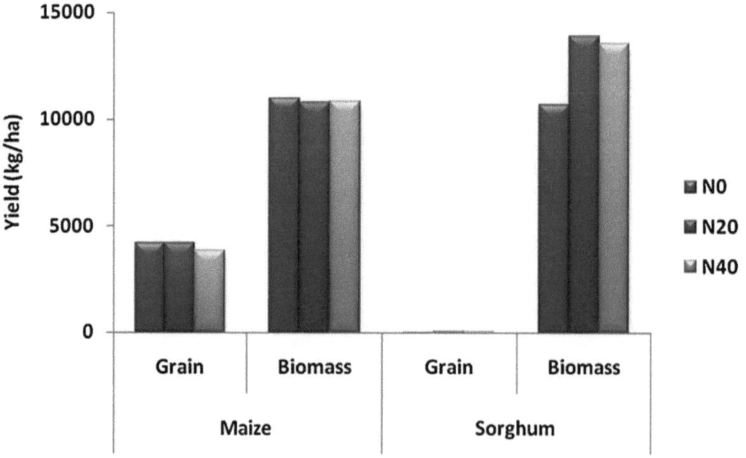

Fig. 8.2 Effect of nitrogen fertilizer rates on maize and sorghum yields (N0 = 0 kg N/ha, N20 = 20 kg N/ha, N40 = 40 kg N/ha)

Sorghum The effect of tillage on mean sorghum yield for LR2012 and LR2013 seasons is presented in Fig. 8.1. Tied ridges increased sorghum biomass yields by 31 % from 11,021 to 14,435 kg/ha while grain yields were similar at 33 and 30 kg/ha for flat and tied ridges tillage, respectively. Significant effect of tillage was observed on biomass yield during the LR2012 season only. The sorghum plots with 20 kg N/ha gave the highest dry matter yield of 13,935 kg/ha and lowest 10,674 kg/ha in plots with 0 kg N/ha, when averaged across the tillage methods (Fig. 8.2). The overall grain yield was negligible due to rot caused by fungal and rust diseases.

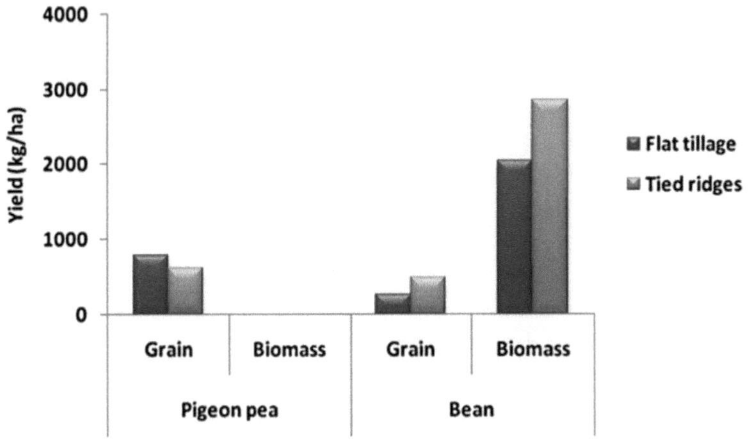

Fig. 8.3 Effect of flat tillage and tied ridges on pigeon pea and common bean yields

Effect of Water Conservation and Plant Population on Crop Yield

Pigeon pea Tillage did not have a significant effect on the pigeon pea grain yield. When grain yield was considered across the plant densities, the grain yield under tied ridges (612 kg/ha) was 21 % less than that under the flat tillage (Fig. 8.4). The medium plant density (33,333 plant/ha) gave the highest grain (719 kg/ha) followed by low (688 kg/ha) and high plant density (687 kg/ha).

Common bean The effect of tillage and plant density on common bean grain and dry matter yields during the LR2012 and SR2012/13 season is given in Figs. 8.3 and 8.4. Tied ridges had a significant effect on bean grain yield improvement during

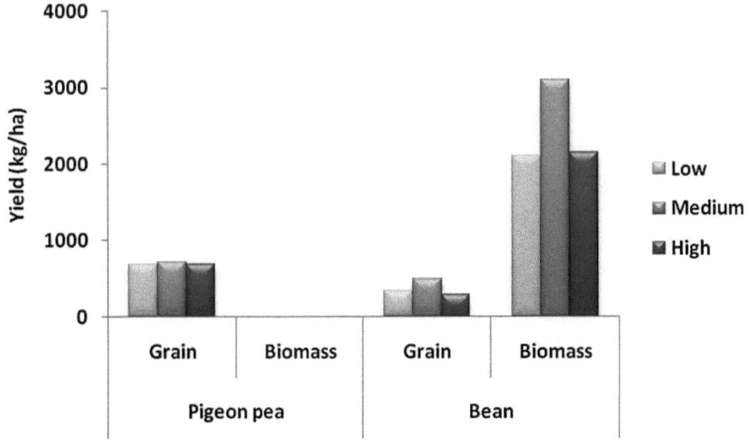

Fig. 8.4 Effect of plant density on pigeon pea and common bean yields

the SR2012 season only. However, average yields for the two seasons showed that the tied ridge tillage yielded higher grain (84 %) and dry matter (38 %) than the flat tillage treatments. Though the bean yield was not significantly affected the plant population in any of the seasons, the medium plant population (148,148 plant/ha) resulted in the greatest grain (491 kg/ha) and dry matter (3,105 kg/ha).

Discussion

The Performance of Crop Varieties

The performance of crop varieties appeared to be affected by the weather conditions. The maturity period for maize, sorghum, pigeon pea and common bean increased by more than 99, 155, 113 and 10 days respectively, when compared to the mean number of days which the crops take to mature in the lower altitudes areas which are warmer and drier (KARI 2006). For example while the optimum temperature for pigeon pea cultivation is 18–38 °C, the mean temperatures during the study period ranged from 13.9 to 19.4 °C. The delayed maturity could largely be attributed to the cold weather conditions in Oljororok. Because of the extended crop cycles, only crop season could be achieved with maize and sorghum in Oljororok unlike in the in lower altitude areas where there are two crop seasons (long rains and short rains) per year.

Though the days to maturity for beans did not change much compared to the other crops, the wet and cold conditions prevailing during long rains seasons encouraged frost and fungal diseases which caused the bean plants to rot; the most affected was the KAT bean 9 bean variety. The highest yields for common bean were obtained during the SR2012 when compared to the long rains season. During the LR2012 (April to September, 2012), the total rainfall and mean temperature were 975 mm and 14.6 °C while the rainfall and mean temperature in SR2012 season (October 2012 to March 2013) were 214 mm and 16.5 °C, respectively. The higher temperatures and drier climatic conditions prevailing in 2012 must have favoured the growth of common bean crop leading to better grain yields. Thus, the short rains seasons are the most suitable for common bean production in Oljororok region.

The most productive crop varieties for maize and pigeon pea at the study site were found to be DK8031 and ICPL 84091, respectively. The yields of 4,938 kg/ha obtained from the DK8031 maize variety are similar to the yields for the highland maize varieties being grown in the area. In general, the maximum legume grain yield of ICPL 84091 pigeon pea variety (881 kg/ha) and GLP24 bean variety (620 kg/ha) obtained in this study were the range of yields production reported under rainfed conditions in Kenya (Katungi et al. 2009). Sorghum grain yield was very low because of fungal and rust disease attack on the crop during the grain filling stage. Kari Mtama 1 sorghum variety was the most affected variety by the diseases. In view of the low yields, the growth of the studied sorghum is not recommended in this region.

Effect of Water Conservation and Nitrogen Fertiliser Application on Crop Yield

Water conservation using tied ridges did not have a significant effect on maize and sorghum yield. Similar findings have been reported by Miriti et al. (2012). When considered across the tillage systems, the highest grain yield of maize and sorghum were obtained in plots with 20 kg N/ha. These results suggest application of 20 kg/ha of N fertilizer was adequate for crop growth at the study site. Although sorghum biomass was high, overall grain yield was negligible (less than 47 kg/ha) due to frost, fungal and rust diseases which caused sorghum heads to rot. The prevalence of the diseases tended to coincide with grain filling stage (November–December).

Effect of Water Conservation and Plant Population on Crop

Though water conservation through ridge tillage resulted in the highest crop yields, the effect was significant on the common bean grain yield only in one season. The medium plant densities for pigeon pea and common bean gave the highest yield, suggesting that plant densities of 148,148 plants/ha for common bean and 33,333 plants/ha for pigeon pea were the most appropriate plant densities. The reduced yield in higher plant densities was probably due to increased competition for soil moisture and nutrients.

Conclusions

Among the tested crop varieties, the most productive varieties at the study site are DK8031, Gadam, ICPL 84091 and GLP 24 for the tested maize, sorghum, pigeon pea and common bean varieties, respectively. The grain production in all sorghum varieties were very low due to frost, fungal and rust diseases. Water conservation using tied ridges did not have a significant effect on the crops yield. The medium plant densities for pigeon pea (33,333 plants/ha) and common bean (148,148 plants/ha) were found to be the most appropriate plant densities for the region. In general, the yield obtained from pigeon pea variety and common bean crops in this study were the range of yields production under rainfed agriculture in Kenya. Based on the results of this study, pigeon pea and common bean can be introduced in the farming systems to improve crop diversity. The best season for growing common bean is during the short rains season when it is warm. In view of the low yields, the growth of the sorghum varieties under study is not recommended.

Acknowledgments This study was supported by the Federal Ministry for Economic Cooperation and Development of Germany through the International Crops Research Institute for the Semi-Arid Tropics (ICRISAT) and Kenya Agricultural Research Institute (KARI).

References

Hay R, Porter J (2006) The physiology of crop yield, 2nd edn, Blackwell Publishing, Singapore

IPCC (2007) Climate Change 2007: Mitigation. In: Metz B, Davidson OR, Bosch PR, Dave R, Meyer LA (eds.), Contribution of working group III to the fourth assessment report of the IPCC. Cambridge University Press, Cambridge, United Kingdom and New York, USA

Jaetzold R, Schmidt H, Hornetz B, Shisanya C (2006) Farm management handbook of Kenya. Natural conditions and farm information, part CB—Central Kenya, 2nd edn, vol II. Ministry of Agriculture, Kenya

KARI (2006) Variety characteristics and production guidelines of traditional food crops. Kenya Agricultural Research Institute. KARI Katumani Research Centre, Nairobi, Kenya, p 44

Katungi E, Farrow A, Chianu J, Sperling L, Beebe S (2009) Common bean in Eastern and Southern Africa: a situation and outlook analysis. Int Centre Trop Agric (ICRISAT) 61:56

Lobell DB, Burke MB, Tebaldi C, Mastrandrea MD, Falcon WP, Naylor RL (2008) Prioritizing climate change adaptation needs for food security in 2030. Science 319:607–610

Miriti JM, Kironchi G, Esilaba AO, Heng LK, Gachene CKK, Mwangi DM (2012) Yield and water use efficiencies for maize and cowpeas as affected by tillage and cropping systems in semi arid eastern Kenya. Agric Water Manage 115:148–155

Mutunga C, Hardee K (2009) Population and reproductive health in national adaptation programmes of action (NAPAs) for climate change. Population action international working paper no. 09-04. UNFCC, p 31

Naab J, Bationo A, Traore PS, Ouattara RT, Vle PLG (2013) African perspectives on climate change and agriculture: impacts, adaptation, and mitigation potential. In: Hillel D, Rosenzweig C (eds) Handbook of climate change and agro ecosystems: global and regional aspects and implications. ICP series. Imperial college Press, London

Tao F, Yokozawa M, Hayashi Y, Lin E (2003) Future climate change, the agricultural water cycle, and agricultural production in China. Agric Ecosyst Environ 95:203–215

Chapter 9
An Assessment of Gender Sensitive Adaptation Options to Climate Change in Smallholder Areas of Zimbabwe, Using Climate Analogue Analysis

Kumbirai Musiyiwa, Walter Leal Filho, Justice Nyamangara and David Harris

Abstract Climate analogues can be used to assess climate–induced risks and adaptation options for smallholder farmers. Surveys were carried out in smallholder areas at two 2050s climate analogue sites to assess smallholder climate-induced risks, farmers' perceptions, and adaptation options, with a gender perspective. Pairs of sites selected had similar annual rainfall totals but differed in mean annual temperature by 2–4 °C. For drier areas Chiredzi was hypothesised to represent Matobo, and for wetter areas Kadoma was hypothesized to represent Mazowe/Goromonzi 2050s climates. Differences in crop management strategies and gender issues vary across sites. At the drier analogue pair, higher proportions of households grew small grains in Chiredzi compared to Matobo. Implications are for increased uptake of small grains, in 2050s climates for Matobo farmers. Gender issues include labour for production and processing of the small grains, against a background of male labour migration. For wetter climates, soil and water management strategies are important options for smallholders. Accesses to draft power, labour, agricultural assets, social and financial capital in differently managed households are important for increasing adoption of effective crop management strategies.

Keywords Adaptation · Analogues · Climate-induced risks · Farming systems · Gender

---

K. Musiyiwa (✉) · J. Nyamangara
International Crops Research Institute for the Semi-Arid Tropics (ICRISAT), Matopos Research Station, P.O. Box 776, Bulawayo, Zimbabwe
e-mail: kmusiyiwa2001@yahoo.com

W. Leal Filho
Faculty of Life Sciences, Hamburg University of Applied Sciences, Lohbruegger Kirchstraße 65, Sector S4, 21033 Hamburg, Germany

D. Harris
Formerly International Crops Research Institute for the Semi-Arid Tropics (ICRISAT), P.O. Box 39063, Nairobi, Kenya

W. Leal Filho et al. (eds.), *Adapting African Agriculture to Climate Change*, Climate Change Management, DOI 10.1007/978-3-319-13000-2_9

Introduction

The impacts of warmer climates by the middle of 21st century, i.e. 2050s on agriculture as a whole and on smallholder production in particular, are projected to be mainly negative (Hulme et al. 2001; Unganai 1996, Christensen et al. 2007). The rural poor, in particular women, are especially vulnerable to the effects of climate change. In Zimbabwe smallholder areas are predominantly semi-arid and rain-fed. Typically smallholders have small land holdings, mostly 2–3 ha of arable land, and are primarily semi-subsistence nature. They are in many instances poorly resourced and associated with low productivity (FAO/WFP 2008, 2010; Chimhowu et al. 2009; Ministry of Agriculture 2007, 2012). The uptake of technologies and interventions to reduce the impacts of bio-physical constraints, which include poor rainfall distribution and soil fertility among other factors, are low due to elements such as costs and skills. Increasing adaptive capacity and options for improved livelihoods in current and future climates for different smallholders requires an integration of methods which include multi-stakeholder processes. Identifying constraints and adaptation options through better understanding of perceived risks and opportunities, and mainstreaming gender issues, specifically women-specific, is important for climate change planning and adaptation.

According to some climate models, temperature increases of about 3 °C by the 2050s and ±5 to 15 % changes in mean annual precipitation are projected for the sub-Saharan African region (Christensen et al. 2007). The potential impacts of warmer temperature and rainfall changes on crop production are projected to be mainly negative (Matarira et al. 1995; Christensen et al. 2007; Thornton et al. 2011). Some direct effects of climate change on crop production include increased evapo-transpiration, water demand and heat stress and shorter growing seasons. Smallholder farmers, who already experience low productivity due to socio-economic and biophysical challenges, are particularly vulnerable to climate change effects. Rain-fed smallholder farmers cope with some of the associated stresses through use of tolerant crops and various soil and water management strategies. These efforts need to be reinforced for current and future climates. For both men and women farmers there is need for enhancement of capital assets i.e. natural, social, human, physical, and financial capital essential for sustainable production. Smallholder systems are heterogeneous. Livelihood options; gender roles, capital assets as well as climate induced-risks differ. Women headed households often achieve lower yields which in part may be due to socio-economic differences (Horrell and Krishnan 2007; Tiruneh et al. 2001). Mainstreaming gender issues in climate and agricultural adaptation require an in-depth understanding of asset ownership at farm levels, of the roles of women and men in the farming systems, as well as benefit sharing of farm profits. Building on existing knowledge and farmer practices is required for progressive adaptation.

Climate analogue analysis can assist in assessing climate-induced risks and their interaction with other capital assets in smallholder production. Spatial temperature analogue models assume that cooler regions will behave like the patterns observed in other regions currently with warmer climates if they were subjected to a climate induced shift (Adams et al. 1998). Burke et al. (2009) demonstrated the existence of analogues for 2050s based upon growing season temperatures for maize, millet, and sorghum in several African countries.

Driven by the need to address the paucity of research on this topic, this paper outlines the use of climate analogues and farmer perceptions in identifying gender sensitive adaptation options in current and future climates for improved livelihoods in some semi-humid (wetter) and semi-arid (drier) smallholder areas in Zimbabwe. The hypothesis of the study was that climate-induced risks and adaptation options will differ by gender in different smallholder areas.

Methodology

The study involved identifying analogue locations in semi-arid and semi-humid areas of Zimbabwe and assessing their perceived climate-induced risks and practices. The project "Adapting agriculture to climate change: developing promising strategies using analogue locations in Eastern and Southern Africa" (CALESA—Climate Analogue Locations in Eastern and Southern Africa) identified climate analogues for semi-arid and semi-humid agricultural areas in Zimbabwe using 30 years climatic data supplied by the Zimbabwe Meteorological Department. The sites selected consist of the wetter analogue pair Mazowe/Goromonzi (reference) and Kadoma (analogue) while the drier analogue pair consists of reference Matobo and its warmer analogue Chiredzi (Table 9.1). Analogue pairs have mean annual temperature difference of at least 2–4 °C.

Table 9.1 Description of selected sites

Surveys districts	Characteristics	Mean annual T (°C)	Mean annual rainfall (mm)	Soil types
Matobo	Cooler/Drier (reference)	18.4	567.1	Greyish brown sands
Chiredzi	Warmer/Drier (analogue)	21.3	541.2	Heavy clays, vertisols, sands, sandy loams
Difference		2.9	−25.9	
Mazowe and Goromonzi	Cooler/Wetter (reference)	18.2	842.9	Greyish brown sands and sandy loams
Kadoma district	Warmer/Wetter (analogue)	21.8	721.7	Greyish brown sands and sandy loams
Difference		3.6	−121.2	

The study involved characterizing farming systems, identifying climate induced–risks, and adaptation options using gender disaggregated data, through household surveys in smallholder areas. To this end, a household survey was conducted at the analogue sites to assess farming practices for the 2010/2011 cropping season. A total of 627 respondents representing different smallholder households were interviewed using semi-structured questionnaires at the study sites. These form the basis of the subsequent analysis presented in the next section.

Results

Characteristics of Households

The proportion of female households for the survey ranged from 26 % (Kadoma) to 43 % (Mazowe/Goromonzi). Greater than 80 % of these were de juri female headed households. De facto female headed households were about 17 % in Chiredzi (warmer, dry), and less than 10 % at the other sites. Higher proportions of female heads were full-time farmers compared to male heads.

Crops Grown at Study Sites During the 2010/2011 Season

In addition to small grains in Chiredzi, main crops grown at the dry sites Matobo and Chiredzi were maize, groundnut and cowpea (Table 9.2). At the wetter sites maize was the main crop.

Soil and Water Management Strategies Used by Households

The identified differences in the uptake and usage of soil and water management practices within and across analogues, are shown in Table 9.3. In general, higher proportions of households from wetter sites have used soil and water management strategies compared to those from the drier sites. At the drier sites, use was generally higher by Matobo households compared to Chiredzi households. The proportion of households that have used conservation agriculture (CA), tied ridges and winter ploughing, was higher in Matobo compared to Chiredzi households; and higher in Kadoma compared to Mazowe/Goromonzi. Soil fertility usage was higher in Matobo, Mazowe/Goromonzi and Kadoma compared to Chiredzi.

Table 9.2 Crop production characteristics at study sites during the 2010/2011 season

	Drier pair			Wetter pair		
	Matobo (cooler)	Chiredzi (warmer)	χ^2	Mazowe and Goromonzi (cooler)	Kadoma (warmer)	χ^2
N	159	165		153	150	
Crops grown by >20 % of households	Maize Groundnut Bambara-nut Cowpea	Maize Sorghum (red, white) Groundnut Bambara-nut Cowpea Pearl millet	N/A	Maize Groundnut	Maize Cotton Groundnut	N/A
Crops grown by <20 % households	Sorghum (white, red) Pearl millet	Cotton Sunflower Sugar bean Soya bean Barley	N/A	Bambara-nut Finger-millet Sugar bean Cowpea Cotton Sunflower Tobacco Rice Okra	Sorghum (white) Cowpea Bambara-nut	N/A
Proportion maize area (%)	79.2	37.6	N/A	89.4	53.1	N/A
Households with small grains (%)	11.9	92.1	208.823***	6.5	4.0	0.947

***Significant at the 0.1 % level
**Significant at the 1 % level
*Significant at the 5 % level

Discussion and Lessons Learned

The hypothesis of the study was that adaptation options will differ by gender and agro-climatic conditions of the smallholder farmers. Results confirm this hypothesis. The results from this study show that characteristics of household heads differ, as well as coping and adaptation strategies for crop production within and across analogues. Higher proportions of de facto female heads at the drier, warmer Chiredzi compared to Matobo and the wetter sites imply higher levels of male labour migration. De-juri female heads are mostly older than the de facto heads.

Table 9.3 Proportion of households who have used soil and water management technologies

Intervention area	Strategy	Drier pair			Wetter pair		
		Matobo (%) hhds[1]	Chiredzi (%) hhds	χ^2	Mazowe (%) hhds	Kadoma (%) hhds	χ^2
1. Soil and water management	Conservation agriculture	53.5	9.1	74.700***	52.9	82.7	30.869***
	Mulching	28.9	15.2	9.7868	60.8	64	0.334
	Contour ridges	47.2	27.7	14.648***	32.7	4.7	38.917***
	Tied ridges	11.9	3.6	7.859**	11.8	21.3	5.033*
	Winter ploughing	10.7	1.2	13.181***	3.3	14.7	12.988**
	Water harvesting	3.8	2.4	0.493	5.9	2.7	1.907
	Pot holing	0.6	5.5	6.304*	6.5	0.7	7.458*
	Multiple weeding	1.3	4.8	3.490	0	0	
	Gulleys	0.6	1.8	0.939	3.3	0	4.984*
2. Soil fertility management	Chemical fertilizer	89.9	4.8	235.584***	99.3	96	3.758
	Animal manure	73	25.5	73.397***	85.6	75.3	5.678
	Compost	71.1	17.6	94.374***	73.2	64	2.980
	Crop rotation	35.2	25.5	3.660	54.2	63.3	2.580

***Significant at the 0.1 % level
**Significant at the 1 % level
*Significant at the 5 % level
[1] households

Higher proportions of female heads who are full time farmers imply higher contribution by rural women in domestic and agricultural production. Differences in household management also influence livelihood sources and access to agricultural resources.

Implications from this study are increased uptake of small grains such as sorghum and millets in Matobo and areas with similar rainfall and temperature characteristics in 2050s. Crop choices are one of the common coping and adaptation strategies employed by rain-fed smallholder farmers (Kurukulasuriya and Mendelsohn 2008). Gender issues for small grain production include labour for production, pest management, harvesting and processing, against a background of male labour migration. At the wetter sites implications are production trends influenced by 2050s market forces. The importance of soil water and soil fertility management particularly at wetter sites is illustrated by high proportions of smallholder farmers who use strategies to conserve soil moisture and increase soil fertility such as use of inorganic and organic forms of fertilizer. The choice of strategies is dependent on, among other factors, soil characteristics, crops grown, capital assets, and easiness to adopt. Nabikolo et al. (2012) showed that adaptation decisions of female heads depended on and were sensitive to more covariates in particular liquid household assets compared to those of male heads. Meanwhile Tazeze et al. (2012) showed gender, age and education level of the household head family size, livestock ownership, agro ecological zones, among other factors influencing adaptation strategies. Gender issues may include, in addition to labour for production, socio-economic factors such as asset ownership, sharing from proceeds after selling crops particularly in male headed.

Conclusions and Recommendations

Current and future projections on climate change and temperature increases in Africa as a whole, and in Zimbabwe in particular, means that there is a pressing need to seek adaptation strategies in agriculture, which may allow farmers to better cope with such changes. Due to a combination of lack of resources, skills and access to technologies, smallholder farmers, especially women, are vulnerable to climate change.

Preliminary results show that climate analogue analysis and involvement of stakeholders such as smallholder farmers through interviews, participatory evaluations can contribute in identifying adaptation options for smallholder farmers in different climates. Gender issues in climate change and agriculture differ and this suggests that differentiated approaches are needed so as to reach male and female farmers. Different preferences for crop management strategies imply gender issues for differently managed households will vary across sites, particularly between the dry analogue pair. In drier areas, implications are for increased uptake of small grains, in 2050s climates. There may be need for increased investment in water management research and development for drier areas. For wetter climates soil and water management strategies are important options for smallholders. At drier sites gender issues include labour for production and processing of the small grains, against a background of male labour migration. At wetter sites access to draft power, labour, agricultural assets, social and financial capital in differently managed households are important for increasing adoption of effective crop management strategies.

Acknowledgments We would like to acknowledge the Deutsche Gesellschaft fur Internationale Zusammenarbeit (GIZ) on behalf of the Federal Republic of Germany who sponsored this study through the "Adapting agriculture to climate change: developing promising strategies using analogue locations in Eastern and Southern Africa"(CALESA—Climate Analogue Locations in Eastern and Southern Africa) Project, ICRISAT—Bulawayo, Zimbabwe for facilitating the field work, the Zimbabwe Meteorological Services Department for providing meteorological data.

References

Adams RM, Hurd BH, Lenhart S, Leary N (1998) Effects of global climate change on agriculture: an interpretative review. Clim Res 11:19–30

Burke MB, Lobell DB, Guarino L (2009) Shifts in African crop climates by 2050, and the implications for crop improvement and genetic resources conservation. Glob Environ Change 19:317–325

Chimhowu A, Bare T, Chiripanhura B, Chitekwe-Biti B, Chung F, Magure T, Mambondiyani L, Manjengwa J, Matshe I, Munemo N, Mtisi S, Nxele M, Sibanda DM (2009) Moving forward in Zimbabwe—reducing poverty and promoting growth. The University of Manchester Brooks World Poverty Institute

Christensen JH, Hewitson B, Busuioc A, Chen A, Gao X, Held I, Jones R, Kolli RK, Kwon WT, Laprise R, Magaña Rueda V, Mearns L, Menéndez CG, Räisänen J, Rinke A, Sarr A, Whetton P (2007) Regional climate projections. In: Solomon S, Qin D, Manning M, Chen Z, Marquis M, Averyt KB, Miller MTHL (eds) Climate change 2007: the physical science basis, contribution of working group i to the fourth assessment report of the intergovernmental panel on climate change. Cambridge University Press, Cambridge, pp 896–901

Food and Agricultural Organization/World Food Program (FAO/WFP) (2008) Special report crop and food supply assessment mission to Zimbabwe. Available at http://www.fao.org/docrep/010/ai469e/ai469e00.htm. Accessed 19 June 2012

Food and Agricultural Organization/World Food Program (FAO/WFP) (2010) Crop and food security assessment mission to Zimbabwe. Available at http://www.polity.org.za/article/faowfp-crop-and-food-security-assessment-mission-to-zimbabwe-august-2010-2010-08-11. Accessed 19 June 2011

Horrell S, Krishnan P (2007) Poverty and productivity in female-headed households in Zimbabwe. J Dev Stud 43(8):80–1351

Hulme M, Doherty R, Ngara T, New M, Lister D (2001) African climate change: 1900–2100. Clim Res 17:145–168

Kurukulasuriya P, Mendelsohn R (2008) Crop switching as a strategy for adapting to climate change. Afr J Agric Resource Econ 2:105–125

Matarira CH, Makadho JM, Mwamuka FC (1995) Zimbabwe: climate change impacts on maize production and adaptive measures for the agricultural sector. In: Ramos-Mane C, Benioff R (eds) Interim report on climate change country studies. US Country Studies Program, Washington, DC

Ministry of Agriculture (2007) Agricultural statistical bulletin. Government of Zimbabwe

Ministry of Agriculture (2012) Mechanization and irrigation development. Second round crop and livestock assessment report. Government of Zimbabwe, Zimbabwe

Nabikolo D, Bashaasha B, Mangheni MN, Majaliwa JGM (2012) Determinants of climate change adaptation among male and female headed farm households in Eastern Uganda. African Crop Sci J 20(2):203–212

Tazeze A, Haji J, Ketema M (2012) Climate change adaptation strategies of smallholder farmers: the case of Babilie district, East Harerghe zone of Oromia regional state of Ethiopia. J Econ Sustain Dev 3:13–14

Tiruneh A, Tesfaye T, Mwangi W, Verkuijl H (2001) Gender differentials in agricultural production and decision-making among smallholders in Ada, Lume and Gimbichu Woredas of the central highlands of Ethiopia. International Maize and Wheat Improvement Center (CIMMYT) and Ethiopian Agricultural Research Organization (EARO), El Baton, Mexico

Thornton PK, Jones PG, Ericksen PJ, Challinor AJ (2011) Agriculture and food systems in sub-Saharan Africa in a 4 C+ world. Philos Trans Royal Soc A 369:117–136

Unganai LS (1996) Historic and future climatic change in Zimbabwe. Clim Res 6:137–145

Chapter 10
Impact of Climate Change and Adaptation Measures Initiated by Farmers

M. Shivamurthy, M.H. Shankara, Rama Radhakrishna and M.G. Chandrakanth

Abstract This paper presents the results from a study conducted to examine adaptation measures initiated by farmers in response to variability and changes in climate and develop a model for climate change adaptation. Drop in rainy days and amount of rainfall concomitant with raise in daily temperature after the year 2000 has been acknowledged widely by farmers. This is corroborated by the actual rainfall and temperatures recorded in the local meteorological observatory. These factors exacerbated groundwater extraction for irrigation and domestic needs. In order to cope with the predicament, farmers shifted to late sowing varieties, dairying and adopted soil and water conservation measures such as farm ponds, ridges and dead-furrows and mulching. Those using conventional irrigation, shifted to drip irrigation, cultivating horticulture crops. Case study of a vulnerable woman farmers indicated that by participating in the groundwater market, the farmer purchased groundwater to cultivate vegetables and keeping three local cows which provided them with steady flow of income from milk. Thus, coping mechanisms were to sustain incomes and resilient due to climate change crisis. To create awareness and understanding of climate change and the challenges for social scientists an integrated model has been proposed by using synergies from program development, program delivery and program impact.

Keywords Adaptation · Capacity building endeavors · Climate change · Coping mechanism · Impact · Perception

M. Shivamurthy (✉) · M.H. Shankara · R. Radhakrishna · M.G. Chandrakanth
Department of Agricultural Extension, University of Agricultural Sciences, GKVK, Bengaluru 560 065, India
e-mail: murudaiah.shivamurthy@gmail.com

W. Leal Filho et al. (eds.), *Adapting African Agriculture to Climate Change*, Climate Change Management, DOI 10.1007/978-3-319-13000-2_10

Introduction

Threats on agriculture and food security are major causalities of climate change in India. Changing climatic conditions leads to substantial influence on human life and environment (World Bank 2007). If one-meter sea level rise were to take place today, it would displace seven million persons in India, (ADB 1995). Increase of CO_2 concentration can lead to an equilibrium warming from 2.0 to 4.8 °C (IPCC 2001). Changes in climatic conditions can be best observed through the extreme rise in temperature, melting of glaciers and sudden rise in sea level. These changes are causing serious problems to human and other forms of life. Poor are likely to be hit hardest by climate change and their capacity to respond to climate change is lowest in developing countries. More than 65 % of total population depends on agriculture in India. Climate change is affecting India and its impacts are erratic monsoons, spread of tropical diseases, raise in sea level, and reduction in availability of surface water and groundwater, floods, droughts, heat waves. In this study, a modest attempt has been made to document the perceptions and adaptations of farmers to climate change using survey and case study methods. Farmers perceived that climate change was apparent after 2000.

Methodology

This study was conducted in the Eastern dry zone of Karnataka state, India, which receives around 500–750 mm of rainfall annually. Considering the variability in rainfall and temperature over the past two decades, 120 farmers belonged to the age group of 35–60 years from three climatically vulnerable districts of Ramanagar, Chickballapur and Tumkur were selected by applying systematic random sampling technique. Data were collected through personal interviews and case study methods by using structured, pretested interview schedule. An attempt was also made to develop a comprehensive model for mitigating the negative effects of climate change. Perception of climate change impacts and adaptation has been defined here as an understanding of climate events and their effects on agriculture and how farmers have prepared and responded to it.

Results and Discussion

Perception of Farmers About Changes in Rainfall and Temperature

Vast majority of farmers (98 %) perceived apparent decrease in rainfall and raise in daily temperature as the two significant changes that are taking place at their locations especially after 2000. There is corroboration between rainfall and temperature data and farmers' perception after 2000 where there was a decrease in the rainfall amount. One significant impact of these changes is increased groundwater extraction.

Coping Mechanisms Adopted by the Farmers Due to Climate Change

Farmer's who are entirely dependent on rainfall generally preferred early maturing variety of finger millet and groundnut before 2000 but shifted to late maturing variety after observing changes in climate (Table 10.1). However, after climate change, farmers preferred late sowing to early sowing in finger millet and groundnut and also in other crops. Farmers have also adopted soil and water conservation measures such as farm pond, ridges and dead furrows and mulching. A sizeable proportion of farmers also had dairy (63 %).

Farmers who have access to groundwater irrigation and cultivating finger millet and other crops have made changes to their systems by shifting to cultivation of horticultural crops and pigeonpea and adoption of drip irrigation system to cope with the changes in climate. Farmers with access to groundwater irrigation and

Table 10.1 Adaptation measures initiated in response to climate change by farmers (n = 120)

Sl. No	Technologies/Practices	After 2000 (% of farmers)	Before 2000
Farmers with groundwater irrigation			
1	(a) Crop shift from finger millet to (b) Crop shift from rice to (c) Crop shift from sericulture to (d) Shift to drip irrigation from conventional flow irrigation	Pigeonpea (65 %) Horticultural crops (65 %) Pigeonpea (85 %) Horticultural crops (15 %) Finger millet (45 %) Horticulture crops (55 %) 36%	No change No change No change 8 %
2	Varietal shift from relatively long to short duration varieties	Finger millet 65 % Rice 60 %	No preference
3	Increase in the number of irrigations given to the crops	Mulberry and rice (100 %)	No change
Rainfed farmers			
1	Preference for late sowing	Finger millet, groundnut (100 %)	No change
2	Soil and water conservation technologies adopted: (a) Farm pond (b) Ridges and dead furrows (c) Mulching	Percent 16 70 33	Percent 7 50 18
3	Integrated Farming Systems	Dairy (63 %), poultry (33 %)	No change

cultivating rice/mulberry silkworms have also followed similar shifts in crops. Those using conventional irrigation, shifted to drip irrigation, as they cultivated horticulture crops. While drip irrigation has been commonly used in irrigating wide spaced tree/horticulture crops, farmers have adopted this even for narrow spaced crops such as tomato, flower crops and banana. This shows that farmers appreciated the economic value of groundwater by adopting drip irrigation to even narrow spaced crops. This is a coping mechanism to manage falling groundwater levels and increasing costs of groundwater arising out of climate change. The cost of groundwater which was Rs. 200 per acre inch before 2000, increased to Rs. 400 per acre inch after 2000, an increase of 100 %. Another coping mechanism observed was shift to relatively short duration varieties from relatively long duration varieties. In addition, farmers have also increased the number of irrigations to meet the increase in croop water requirements for the increase in day temperature.

Micro-level Implications of Climate Change

The case study of a woman farmer who has adapted to climate change indicated that the farmer by participating in the groundwater market, purchased groundwater to cultivate vegetables as also rear three local milch cows (Table 10.2). Further the farmer increased her flock of sheep from 8 to 40, thereby enhancing the income from sheep to Rs. 60,000, an increase of 300 %. By participating in the groundwater market, the farmer purchased groundwater and cultivated vegetables. The income increased from Rs. 30,000 to Rs. 151,000, an increase 400 %. Thus, the woman farmer became economically empowered due to her participation in the water

Table 10.2 Socio economic characteristics of case study woman farmer

Sl. No	Particulars	Before climate change		After climate change	
		Holding size and enterprise combinations	Annual income (Rs.)	Holding size and enterprise combination	Annual income (Rs.)
1.	Land holding	2 acre of rainfed land cultivating finger millet, sesamum)	15,000	2 acres (1 acre of rainfed land for finger millet + 1 acre of land cultivating vegetables using purchased water)	80,000
2.	Sheep	8	15,000/-	40	60,000
3.	Local cow	–	–	3	5,000
4.	Vegetable selling activity	–	–		5,000
Total (Rupees, 1 USD = 58 rupees at the time of study)			30,000/-		151,000

market from where she purchased water for irrigation. Drilling irrigation well is not only expensive but also risky venture, given the high probability of well failure due to climate change.

The farmer cultivated horticultural crops in place of field crops. By discontinuing field crops due to non-remunerative prices in the market, the farmer also focused on milch cows, sheep and backyard poultry. Farmer abandoned goat rearing as goats were voracious feeders of tree fodder which the farmer could not afford. Coping mechanisms of climate change were to sustain income in the wake of climate change crisis.

Effects of Climate Change on Health Aspects of Farm Women

Regarding ill effects of climate change, the farm women expressed sun scorch due to high daily temperature after 2000. This resulted in weakness, rashes on the skin, body pain and headache and led to increased frequency in consulting the doctors, increased medical expenses and difficulty in working in the field (Table 10.3).

Synergies of Climate Change

The efforts towards understanding the negative effects of climate change have enhanced our understanding of climate change and its effect on natural resources and environment. However, awareness and understanding of climate change has been limited to the level of scientists and has not reached grassroots level. Challenge for social scientists and educators, in agriculture sector is to create awareness of the effects of climate change on people, especially farmers, community leaders, policy makers and general public. The following model helps to appreciate and

Table 10.3 Effects of climate change (high temperature and sunshine)

Sl.No	Adverse health aspects	Severity before 2005	Severity after 2005	Impact of adverse health effects and coping mechanism
1	Sun scorch	Tolerable	Severe	Consulting doctors–once or twice a year and medical expenses were affordable before 2000. However, medical expenses have tripled after 2000 Weakness and laxity for working in the fields due to high temperature and sunshine
2	Feeling weakness	Not felt	Severe	
3	Rashes on the skin	Less noticed	More	
4	Headache & body pain	Not much felt	Frequently experienced	

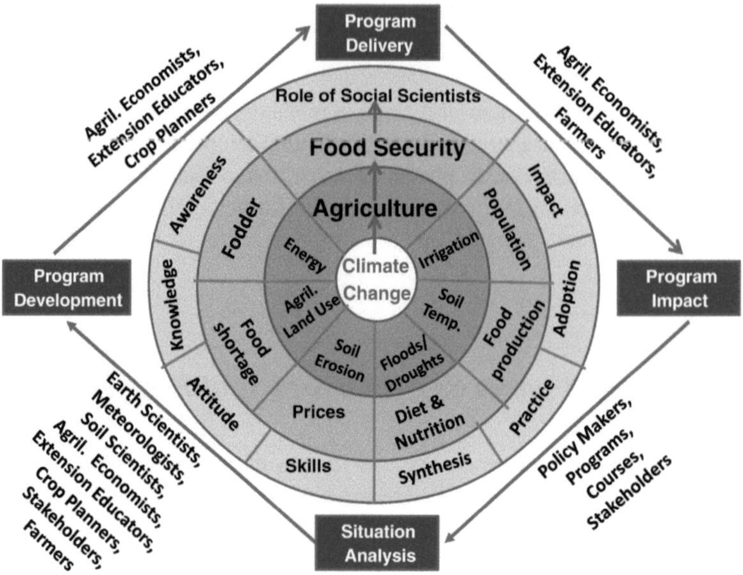

Fig. 10.1 Synergies of climate change

understand the comprehensive strategies (Fig. 10.1) to address climate change issues.

The strategy involves four phases. In the **situation analysis phase**, it is important to take stock of the existing body of knowledge generated in relation to climate change. In this phase, collaborative and integrated approaches to understand climate change issues are discussed. Scientists, meteorologists, crop and soil scientists, extension educators, agricultural economists, sociologists, farmers, and other stakeholders should come together to discuss the climate change issues and offer solutions for the same. The rationale in this phase is to bring together the collective expertise of all and to translate the scholarship of research on climate change to the scholarship of outreach.

In the **program development phase**, extension educators and economists, equipped with research-based knowledge and information, will develop educational programs to be delivered to the affected people–farmers, local leaders, policy makers, and government officials. The intent of the program development is to create awareness among farmers on many facts of climate change and its effects on agriculture, including food security and the environment. The role social scientist, especially extension educators, sociologists and economists is critical. The challenge is how to put together an educational program on climate change to create awareness, knowledge acquisition, skill development, attitude change, and ultimately adoption.

The third phase is the actual **delivery of climate change educational program** in areas where climate change has adversely affected in terms of crop growth, land use, soil erosion, etc. in addition, the program planners need to identify the location, the target population, learning experiences to be provided, and informational materials to be developed and shared. Then they have to deliver the program using a variety of methods- lecture, discussion, demonstration, field visits, guest lectures etc. Understanding the needs of affected people and their educational level is critical in this phase for the successful delivery of the program.

In the final phase, extension educators and economists have to **document evidences on impact of the program** in reducing the effects of climate change either at the micro level or at the macro level. In this level, the changes occurring in farmer's field change in cropping pattern, land use, and other aspects are to be documented. Besides, Changes in farmers' knowledge, attitude, and skills along with alternative strategies to reduce climate change effects that are economically viable, technically feasible, and socio-culturally acceptable are developed.

Conclusions

This study focused on the perceptions, awareness and coping mechanisms adopted by farmers in the wake of climate change, in eastern dry zone of Karnataka, India. A vast majority of the farmers expressed that the daily temperature increase and fall in monsoonal rainfall have occurred conjointly. Accordingly, farmers have shifted to late planting of short duration crops and varieties. Those using groundwater irrigation have shifted to drip irrigation for narrow spaced horticulture crops, which in fact was largely used for broad spaced crops. Farmers without irrigation are also switching over to late planting of short duration crops and varieties, along with adoption of soil and moisture conservation practices such as farm ponds and mulching. Purchasing groundwater and cultivating horticulture crops are other coping mechanisms adopted. These farmers are also rearing milch cows and sheep to benefit from the growing demand for milk and meat. The coping mechanisms have increased the incomes of farmers substantially (>100 %). However, groundwater farmers are not effectively recharging their irrigation wells, which is a further cause of worry. Similarly market reforms breaking the barriers of asymmetric information are yet to come to the fore to help farmers who are in the imminent threat of distress sales.

Thus, the challenge is the need for a coordinated, collaborative and integrated approach to challenges posed by climate change. Scientists, regardless of being labeled social or pure/fundamental, must bridge the gap, link research to practice through education, and become proactive to challenges posed by climate change. This calls for an integrated, collaborative approach to study the effects of climate change. Further, this needs development and delivery of educational programs at grassroots level to create awareness and understanding of climate change issues. A systematic review and assessment of curricula at the primary and secondary school levels to include climate change as a subject of study is a dire necessity.

Reference

IPCC (2001) Climate change: impacts, adaptation and vulnerability. In: McCarthy JJ, Canziani OF, Leary NA, Dokken DJ, White KS (ed) Third assessment report. Cambridge University Press, Cambridge

Chapter 11
In Situ Soil Moisture Conservation: Utilization and Management of Rainwater for Crop Production

P. Kathuli and J.K. Itabari

Abstract The salient results of in situ soil water conservation technologies that have been extensively tested and found suitable for increasing soil moisture for increased land productivity in the arid and semi-arid lands (ASALs) of eastern Kenya are reviewed in this paper. The technologies reviewed are *fanya juu* terraces, contour bunds, negarims, trapezoidal bunds, on-farm micro-catchments, *Zai* pits, *tumbukiza*, tied ridges, deep tillage and sub-soiling and ripping. These technologies hold rainwater on the soil surface thereby allowing it more infiltration time leading to enhanced soil moisture status, which would not be attained in the absence of these interventions. *Zai* pits, *tumbukiza* and deep tillage when used together with soil fertility improvement can increase crop yields by 4–10 times when compared to other similar fields cultivated conventionally. When tied-ridging tillage is used together with fertilizer, manure or their combination it can increase crop yields by 100–300 %. Sub-soiling and ripping increases crop yields by 50–100 % when used together with soil fertility improvement. Micro-catchment technology at 1:1 and 2:1 catchment to cultivated land ratio can increase crop yields, but is not practised due to land limitation. Use of fertilizers and or manures with in situ soil moisture conservation leads to improved water use efficiency by crops planted in the semi-arid eastern Kenya. It is, therefore, proposed that in situ rainwater conservation technologies should be an integral part of the farming systems for increased soil moisture conservation, crop production and food security in the semi-arid Eastern Kenya.

Keywords In situ rain water harvesting · Soil fertility improvement · Crop yields · Semi-arid eastern Kenya

P. Kathuli (✉) · J.K. Itabari
KARI-Katumani, P.O. Box 340-90100, Machakos, Kenya
e-mail: peterkathuli@yahoo.com

W. Leal Filho et al. (eds.), *Adapting African Agriculture to Climate Change*,
Climate Change Management, DOI 10.1007/978-3-319-13000-2_11

Introduction

In situ soil moisture conservation entails capturing rain water and retaining it in soil for in situ plant utilization for growth and increase in grain and biomass yield. This is achieved through rain water harvesting on-farm where crops or fodder are planted to benefit from the conserved rainwater. It reduces rainwater loss through runoff, thereby increasing the amount of rain water that will be useful for crop and fodder production through increased rain water infiltration time and hence increase in food security (Itabari and Wamuongo 2003).

Arid and semi-arid lands (ASALs) constitute 83 % of Kenya's land mass and support over 25 % of the human population. The problem of food insecurity in this region is mostly due to declining land productivity (Itabari et al. 2011), which is mainly attributed to inadequate soil moisture to support crop and fodder production. The rain water balance in the ASALs of eastern Kenya shows that 60–70 % of the rainwater is lost as runoff, leaving only 25–30 % for crop and fodder production (Kilewe 1987). Most soils in this region have developed hardpans at a depth of around 10 cm, which limit rain water infiltration. This leads to rain water loss through runoff, which carries away top fertile soil required for healthy crop growth. The soils in the region are susceptible to hardpan formation due to their inherent low organic matter and weak soil aggregate stability. The soil aggregates are broken down to fine aggregates, which join and form a seal of impermeable layer with time. The soils become susceptible to erosion (Kilewe 1987), losing rain water due to decreased infiltration as the bulk density of the few centimeters depth of the top soil increases, thereby limiting infiltration. This has been observed in semi-arid Makueni District (Miriti et al. 2009) where the top 0–15 cm soil had higher bulk density than the soils below this depth. An appropriate soil management technique is required for soils in the ASALs because of these inherent properties.

Also, the low and erratic 100–700 mm annual rainfall in the region means that crops and fodder production can only be achieved if in situ form of rain water harvesting is carried out for rainfed cropping system to increase soil moisture amounts and retention for crop production. In situ rain water harvesting can increase crop production by more than 40 % (Kathuli et al. 2010; Gichangi et al. 2007; Itabari et al. 2003) provided other crop and fodder agronomic practices are carried out. In other arid and semi-arid regions of eastern Africa, in situ rain water harvesting increases rain water productivity or efficiency from 1–1.5 to 3–4.5 kg/mm rain (Steiner and Rockstrom 2003).

Crop water utilization and efficient use of rain water is based on the principle that, the loss of rain water from the moment it reaches the soil surface and ensuring that its utilization by crops is as efficient as possible. The amount of water available for utilization by a crop is expressed as $T_{crop(mm)} = P\text{-}R\text{-}D\text{-}E\text{-}T_{weeds} - \Delta_s$ (Itabari 1999) where T = transpiration, P = precipitation, R = runoff, D = drainage, E = evaporation from the soil surface and Δ_s is change in soil water stored within the rooting zone.

The equation shows that, in order to have good rain water utilization by plants, T_{crop} must be optimized. This can only happen if R, D and E are minimized through soil management. The water utilization efficiency is calculated as $\mathrm{WUE} = \mathrm{Crop\,Yield}/\mathrm{T_{crop}} = \mathrm{kg/mm\,rainfall}$

The objective of this paper was to review the salient results of in situ soil water conservation technologies that have been extensively tested and found suitable for increasing soil moisture for increased land productivity in the ASALs of eastern Kenya and particularly in areas found within agro-ecological zone IV and V.

Materials and Methods

Secondary data from research conducted in the ASALs of eastern Kenya was collected from journals, conference and workshop proceedings and technical reports. The data represent the status of in situ soil moisture conservation in the ASALs of eastern Kenya. These lands are found in agro climatic zone (ACZ) IV, V, VI and VII. Agro-climatic zone IV has mainly low to medium altitudes, ranging from 1,300 to 1,800 m above sea level. The mean annual temperatures here range from 18 to 21 °C. Agro-climatic zone V falls within 1,800–1,300 m above sea level, and the mean annual temperatures range from 21 to 24 °C. For agro-climatic zones VI and VII, the mean daily maximum temperature varies from 32 to 37 °C in the cooler and hotter months, respectively. The mean minimum temperature is about 22–23 °C, giving a diurnal variation of 10–15 °C (Sombroek et al. 1982; Jaetzold and Schmidt 1983).

The annual rainfall ranges from 500 to 800 mm, and is erratic in amount and distribution. It is bimodal in the areas along the coastal hinterland and in the areas around the eastern slopes of the central highlands, and unimodal in the areas to the West of the central highlands. In the areas where it is bimodal, it is almost evenly distributed between the 'long rains (March–May) with a peak in April, and the 'short' rains (October–December) with a peak in November. The rates of evaporation are generally high due to the high daytime temperatures. Rates of up to 8.2 mm per day have been recorded at Katumani, which lies in ACZ IV (Stewart and Hash 1982), and over 3,000 mm year^{-1} for agro-climatic zones VI and VII (Sombroek et al. 1982; Jaetzold and Schmidt 1983).

The predominant soil types are Luvisols, Acrisols and Vertisols. There are other soil types, but they are of less significant in terms of the agricultural area they occupy. Their texture ranges from sandy loam to loamy sand with a tendency to harden when dry, but are friable when wet. They are deep and well-drained in the wetter areas, but tend to be shallow in the drier areas due to the presence of petroplinthite horizons. They have low organic matter content (less than 1 % C), mainly due to the poor growth of the natural and human-modified vegetation and removal of crop residues for livestock feed. They have a low water-holding capacity, are generally medium to slightly acidic (pH 5.0–6.5) in the surface horizons, poor structural development, are highly erodible and prone to surface sealing and capping through the energies of high-intensity rainfall and solar radiation (Muchena 1975).

Results

Some of past researched technologies on in situ rainwater conservation technologies include the tied and open-ridges, *zai* pits, semi-circular hoops, stone bunds, negarims, contour bunds, terraces, trapezoidal bunds, micro-catchments (Itabari et al. 2004), deep tillage and sub-soiling/ripping (Kathuli et al. 2010; Miriti et al. 2009; Steiner and Rockstrom 2003). These technologies allow rain water retention for a prolonged duration on the soil surface for increased infiltration and retention and better rain water use efficiency (Steiner and Rockstrom 2003; Itabari 1999). The following is a description of each of these in situ rain water conservation and utilization with highlights on successes and constraints.

Fanya Juu Terraces

Fanya juu terraces are constructed by heaping soil up-slope to make an embankment which forms a runoff barrier leaving a trench used for retaining or collecting runoff. The canal is 0.6 m deep and 0.6 m wide. The soil embankment is about 0.7 m from the surface. Runoff from external catchments is led into the canals for retention to allow more time for water to infiltrate into soil. Crops like bananas, pawpaws and citrus can be grown in the ditches. This technique is recommended for areas with slopes greater than 5 % (Plate 11.1).

Plate 11.1 Suitable land topography for construction of fanya juu terraces for in situ conservation of rain water for crop and fodder production in ASALs of eastern Kenya

Contour Bunds

Small earth, stone or trash lines embankments are constructed along a contour line to form an embankment. The embankments trap rain water flowing down the slope and retain it behind the bunds. The area behind the bunds can be levelled to ensure homogeneous infiltration. The spacing of the contour bunds depends on slope and soil type. Land on steep slopes will require closer contour bunds. Catchment to cultivated area ratio should be 2:1. A successful sorghum crop has been achieved with 270 mm rainfall using this technology (Itabari and Wamuongo 2003).

Semi-circular Bunds (Hoops)

These are semi-circular earth embankments with tips of the bunds on the contour. Water is collected within the hoop from the area above it and confined to the depth defined by the height of the bund and position of the tips. Excess water is discharged around the tips. An illustration on use of this technology is shown in Plate 11.2.

Negarims

These are small V-shaped embankments, with the apex at the lowest point. Water is collected from the V-shaped basin and stored in the soil profile at the apex. This technique is good for the establishment of trees and shrubs. Catchment area ranges

Plate 11.2 Semi-circular bund ready for grass reseeding in semi-arid Taveta sub-county

Plate 11.3 Farmers in semi-arid Makueni County demonstrating negarim making for tree crop establishment (Kathuli and Mweki 2012)

from 16 m^2 in agro-ecological zone (AEZ) IV to 1,000 m^2 in AEZ V. The soil embankment is 15–20 cm for water collection while the apex basin is 40 cm deep for water storage (Plate 11.3).

Tied Ridges

These are made to increase surface storage and to allow more time for rainwater to infiltrate the soil. Oxen made furrows are manually tied at 3–5 m intervals. The lower furrow is tied starting from the point between the above tied furrows such that tying is not perpendicular to prevent possible erosion in the farm to give a pattern similar to house construction using bricks. The cross ties are usually of lower height than the furrow so that if they fill, the overflow is along the furrow but not down the slope. This technology is recommended for land having a slope greater than 2 % so that the furrows retain rain water that would be lost as runoff if the structures were not in place. Effect of tied and open ridging technology of rainwater harvesting on mean maize grain yields across various sites in semi-arid eastern Kenya is shown in Tables 11.1, 11.2 and 11.3.

Data from Gichangi et al. (2007) on maize grain yields across various sites in semi-arid of eastern Kenya (Table 11.1) show that tied-ridging with manure, manure and fertilizer application can increased crop yields from 100–359 %.

Similar results were obtained in Mwala during the short rains of 2007 although the maize total dry matter (TDM) yields were not significantly different from conventionally cultivated fields (Table 11.2).

Tied-ridging with fertilizer use significantly ($P = 0.05$) out-yielded plots with conventional cultivation during the 1995 short rains (Table 11.3).

Table 11.1 Effect of tied-ridge rain water harvesting on mean maize grain yields across various sites (kiomo, masii, mavuria and kwa vonza) in semi-arid eastern Kenya

Treatments	Mean grain yield (kg/ha)		Percentage of yield increase + water harvesting	Percentage of increase − water harvesting
	Tied ridging	Open ridging		
0 t/ha FYM	655.0	483.0		
10 t/ha FYM	1319.4	788.0	101.4	63.1
20 t/ha FYM	1866.9	1284.0	185.0	165.8
20 kg N/ha	1466.9	1167.0	123.8	141.6
20 kg N, 20 kg P_2O_5	2035.0	1603.0	256.5	231.9
10 t FYM, 20 kg N	2536.8	1784.0	287.3	269.4
10 t FYM, 20 kg N, 20 kg P_2O_5	3007.0	2,155	359.1	346.2
$Lsd_{0.05}$	407.6	407.6		

FYM farmyard manure
Source Gichangi et al. (2007)

Table 11.2 Effect of tied ridging on TDM yield (kg/ha) of maize in mwala (aez 4), yatta and kitui (aez 5) during the 2007 short rains

District	Farmers	Tied ridging	Open furrows
Mwala	P. Kyululi	1,414	1,393
	A Musyoka	3,814	3,384
Yatta	T. Muthama	308	254
	M. Ndolo	193	125
Kitui	M. Mwava	204	157
Mean yields		1186.6	1062.6

Source Kathuli et al. (2010)

Table 11.3 Effect of tied-ridging and fertilizer on grain yield and water-use efficiency (wue) of sorghum at masinga during the 1995 short and long rains

Treatments/seasons	Grain yield (kg ha^{-1})	ET	WUE (kg ha^{-1} mm^{-1})
Short rains			
Flat cultivated − fertilizer	190b	299.0	0.64
Flat cultivated + fertilizer	380b	299.2	1.27
Tied ridging − fertilizer	360b	297.8	1.21
Tied ridging + fertilizer	820a	300.5	2.73
Long rains			
Flat cultivated − fertilizer	80c	276.09	0.29
Flat cultivated + fertilizer	350abc	276.86	1.26
Tied ridging − fertilizer	310bc	275.53	1.13
Tied ridging + fertilizer	1030a	276.97	3.75

Values in the same column followed by the same letter are no significantly different ($P = 0.05$) level
Source Itabari (1999)

Plate 11.4 Illustration of a trapezoidal bund in semi-arid Taveta sub-county

Trapezoidal Bunds

These are bunds with trapezoidal-shaped earth embankments (Plate 11.4).

Tips of embankments are placed on the contour. The embankment top is level and higher than the ground level at the tips. Water flowing down-slope is trapped and retained behind the bund up to the level of the tips and any excess overflows around the tips into other bunds in the system or natural drainage course. The size of enclosure depends on slope and may vary from 0.1 to 1.0 ha. Embankment base width varies from 2.6 to 5.8 m. ratio of catchment to cultivated area varies from 1:1 to 5:1 depending on rainfall regime, soil properties and crop water requirements. This technology can increase sorghum yields by 30–90 % in ASALs of eastern Kenya (Itabari and Wamuongo 2003).

Zai Pits

These are small planting pits 30 cm in diameter and 15–20 cm deep. Manure or compost is placed at the bottom of the pit and mixed with soil before planting. During digging, the soil is thrown down-slope to form a small embankment. The pits are made at a spacing on one meter row to row and pit to pit can be 30 cm. the pits should not be perpendicular to each other to avoid possible erosion in case of a heavy rainfall. These are useful for establishing vegetables and field crops. Some results obtained from *zai* pits are shown in Table 11.4.

Sub-soiling and Ripping

This involves use of a sub-soiler and a ripper which are drawn by animals. The sub-soiler is made of a round steel metal with a sharp end with an attachment to the ox plough. It is adjustable and penetrates the soil to a depth of 15 cm breaking soil

Table 11.4 Effect of tillage and fertilizer on grain yield and water-use efficiency (wue) of sorghum at masinga during the 1995 short and long rains

Treatments/seasons	Grain yield (kg/ha)	ET	WUE (kg/ha/mm)
Short rains			
Zai pitting − fertilizer	850a	297.9	2.85
Zai pitting + fertilizer	1010a	298.8	3.38
Long rains			
Zai pitting − fertilizer	900ab	275.10	3.27
Zai pitting + fertilizer	780ab	275.99	2.83

Values in the same column followed by the same letter are no significantly different at $P \leq 0.05$ level

Source Itabari (1999)

hardpan, usually formed at 10 cm soil depth due to continuous cultivation at this depth. The ripper opens the narrow furrow left by the sub-soiler to 8 cm wide ready for planting. The seeds are placed at the bottom of the furrow and covered lightly. The furrow traps rain water and holds it for some time as it infiltrates the soil. Sub-soiling/ripping increases crop production by 40–60 % in the ASALs (Steiner and Rockstrom 2003; Mwangi et al. 2005). In the absence of in situ rainwater harvesting, these lands lose 60–75 % of rain water through surface runoff due to their topographic features and inherent soil properties (Kilewe 1987). Combination of this in situ rainwater harvesting technologies and integrated soil fertility improvement technologies has led to increased crop yields (Mwangi et al. 2005, Kathuli et al. 2010). Similar results were obtained by use of sub-soiling/ripping in situ water harvesting technologies with sorghum, sorghum-cowpea intercrop and sorghum rotation with and without manure which resulted in increased sorghum yields (from 0.36 to 1.96 t/ha) where manure was applied in Makueni county (Kitinya et al. 2011). Results of on-farm trials involving the use of sub-soiling and ripping technology for in situ rain-water harvesting for soil moisture conservation on total dry matter yield and grain yields are shown in Tables 11.5 and 11.6, respectively.

Tumbukiza

These are planting pits, 60 cm wide and 60 cm deep. They are modifications of *Zai* pits. The top 0–20 cm soil is mixed with manure or compost prior to planting. 5–7 maize seeds can be planted per pit. They are spaced 100 cm row to row and 75 cm pit to pit and can also be used to establish fodder crops. This technology is illustrated below (Plate 11.5).

Table 11.5 Sub-soiled/ripped and conventionally tilled farms during the 2007 short rains and the 2009 long rains in mwala

Farmers name	District	Season	Total dry matter yield (kg/ha)		Yield increase (%)
			Treatments		
			Sub-soiling/ ripping	Conventional tillage	
A Muli	Mwala	SR 2007	239	102	134
Tabitha	Kitui	SR 2007	50	25	100
Kalumu	Kitui	SR 2007	350	167	109
Mean			213	98	117
Alize Musyoka	Mwala	LR 2009	189.8	222.2	
B Muoki	Mwala	LR 2009	766	570.0	34
M Kiingu	Mwala	LR 2009	283	438	
A Muli	Mwala	LR 2009	590	189	212
Mean			457.2	355	29
Lsd (5 % level)			392.3	392.3	
CV%			47.4	47.4	
s.e.d			123.3	123.3	
($P \leq 0.05$)			NS	NS	

Source Kathuli et al. (2010)

Table 11.6 Effect of integrated soil fertility and tillage methods on maize grain yields (kg/ha) in mwala aez-4 and yatta aez-5 during the 2009 short rains and kitui aez-5 during the 2007 short rains

Site	Soil fertility management	Treatments		Yield change (kg/ha) (%)
		Sub-soiling /ripping	Conventional tillage	
Kyasioni (AEZ-5) Yatta	5 t/ha FYM	1,530	1,630	−6
	5 t/ha FYM + 20 kg N/ha	1,710	1,080	+58
	(20 kg N + 20 kg P_2O_5)/ha	3,710	933	+298
	5 t/ha FYM +(10 kg N + 10 kg P_2O_5)/ha	2,210	1,340	+65
	Mean	2,290	1,246	+84
Kyawango (AEZ-4) Mwala	5 t/ha FYM	2,430	1,680	+45
	5 t/ha FYM + 20 kg N/ha	2,950	1,810	+63
	(20 kg N + 20 kg P_2O_5)/ha	2,220	2,020	+10
	5 t/ha FYM + (10 kg N + 10 kg P_2O_5)/ha	3,370	2,450	+38
	Mean	2743	1,990	+38
Kauwi (AEZ-5) Kitui	5 t/ha FYM	50	25	+100
	5 t/ha FYM	350	167	+109
	Mean	200	96	+108

Source Kathuli et al. (2010)

Plate 11.5 Napier planted in conventionally cultivated field and in tumbukiza planting pits at Katumani Machakos

Table 11.7 Sorghum and cowpea grain yields from a runoff harvesting trial at katiorin (1981) with different tillage treatments

Treatments	Sorghum(kg/ha)		kg/ha
Treatments	First harvest	Ratoon harvest	Cowpea
Impounded plot, deep tillage	420	595	70
Impounded plot, zero tillage	120	–	–
3 m contour ridges hoop, zero tillage	410	900	130
Control plot deep tillage	60	325	20

Source Imbira (1989)

Deep Tillage

Once soils are deep ploughed, rain water infiltration is increased. Rain water storage by the soil is also increased and yield can increase 3 times in comparison to land tilled conventionally (Imbira 1989). Field data on effect of deep tillage on sorghum grain yield from dry land Katiorin during the 1981 long and short rains are shown in Table 11.7.

Deep tillage increased sorghum yield by 3.5 times when compared to a farm with same management and less deep tillage. Both plots had rainwater harvesting with one plot being deep-tilled.

Micro-catchments

This involves capturing runoff from upper part and collecting it in adjacent lower part of the farm. The soil in the lower farm is cultivated to increase water infiltration. The ratio of the catchment to cultivated area usually varies from 1:1 to 5:1 depending on the rainfall regime, soil properties and crop water requirements. Yield increase of 30–90 % has been obtained using this technology in the semi-arid Baringo sub-county on a sorghum crop (Imbira 1989) (Table 11.8).

Table 11.8 Yield of sorghum from trial plots using on-farm external catchment systems during the 1982–1983 long rains

Plot	Year	Catchment: cultivated area ratio	Experimental plot yield (kg/ha)	Control plot yield (kg/ha)	Percentage of yield increase (%)
Katiori	1982	2:1	775	135	474
Marigat	1983	1:1	540	10	5,300

Source Imbira (1989)

Discussion

The technologies reviewed here allow rain water retention for a prolonged duration on the soil surface for increased infiltration and retention and better rain water use efficiency (Steiner and Rockstrom 2003; Itabari 1999). In absence of these technologies, the farms would be losing about 70 % of rainwater to runoff and leaving only 25–30 % for crop or fodder production (Kilewe 1987).

Fanya juu terraces are constructed on land with slope ranging from 2–22 % (Itabari et al. 2011). These structures form a runoff barrier which collects runoff rainwater and eroded soil. The water collected spreads back in the terrace and is retained for a longer time allowing infiltration and raising soil moisture for fodder and crop production. The effects of fanya juu terrace (Plate 11.1) indicates that, the trapped rain water would have been lost in the absence of these structures thus limiting crop productivity. The in situ conserved rain water will increase crop and fodder productivity in ASALs where water is the most limiting crop and fodder production constraint (Itabari et al. 2011). However, these structure require periodic maintenance to increase the embankment height as more soil is removed from upper side of the terrace and deposited in the lower side to form a bench terrace. There are reports of increased crop productivity in those farms where terraces are periodically maintained (Itabari et al. 2011).

The contour bunds embankments trap rain water flowing down the slope and retain it behind the bunds. The area behind the bunds can be levelled to ensure homogeneous infiltration. This technology concentrates rain water in a smaller area for cultivation of early maturing crops. The technology has led to a satisfactory sorghum crop with rainfall of 270 mm using a catchment to cultivated area ratio of 2:1. However adoption of this technology is not wide spread (Itabari and Wamuongo 2003) due to land scarcity.

Semi-circular bunds (hoops) are common is ASALs of Turkana and Baringo counties. The bunds are used to capture rainwater that would be lost as runoff in the absence of these structures due to land topography. The rainwater is retained in the structures allowing longer infiltration time. They are used to rehabilitate degraded lands (Kitheka et al. 1995) in ASALs of Kenya. Restoration of productivity is achieved within three seasons. The bunds are used for the reseeding of grass, fodder, shrubs and can be used to grow early maturing crops like cowpea and green grams (Plate 11.2). Adoption of the technique has been hampered by labour involved in construction of the structures.

Negarims are suitable in establishment of trees or tree crops (Itabari et al. 2011). Under very low rainfall, the runoff is concentrated into a planting pit thus increasing soil moisture for tree crop establishment and growth (Kathuli and Mweki 2012). These structures improve fruit tree establishment by 60 % leading to increased yields and farm income. The structures are recommended in areas with 300–700 mm annual rainfall and with 1–5 % slope (Critchley et al. 1991).

The effect of tied-ridging with and without fertilizer use on sorghum grain yield and water use efficiency in semi-arid lands of Masinga in eastern Kenya during the 1995 short and long rains (Table 11.3) shows that, tied-ridging plus fertilizer significantly ($p = 0.05$) increased sorghum grain yield in both seasons. Rain water use efficiency was similarly enhanced by combination of tied-ridging and fertilizer use. Evapotranspiration (ET) remained fairly constant within the seasons. Tied-ridging has been shown to increase total dry matter of maize by less than 1 % in very poor seasons (Table 11.2) (Kathuli et al. 2010) while in good rainfall seasons, sorghum yields increased by 3–5 % across short rains and long rains (Table 11.3) (Imbira 1989) and maize yields increased by 63–340 % due to tied ridging with integrated soil fertility management (Gichangi et al. 2007). Tied ridging allows rain water to be conserved in situ as it infiltrates the soil. The prolonged time it is retained on furrows allows increased infiltration and hence increased soil water which is used by crops. The water use efficiency by crops is increased with addition of manures and fertilizers (Tables 11.1, 11.3 and 11.4).

Similarly, *Zai* pitting significantly ($p \leq 0.05$) increased sorghum grain yields and improved sorghum water use efficiency in Masinga, Machakos County during the 1995 short rains. Sorghum yields increased 4 times with fertilizer and 2 times without fertilizer use on *Zai* pits. *Zai* pitting without fertilizer application significantly ($p = 0.05$) increased sorghum grain yields by 10 times over the plot without *Zai* pitting and with no fertilizer during the same period (Tables 11.3 and 11.4). This is attributed to limited rainfall with same fertilizer during the short rains. *Zai* pits can increase crop yields by a bigger margin with soil fertility improvement. This is because rainwater is concentrated into a smaller area increasing soil water per unit soil volume thus providing adequate soil moisture that favor crop growth and yield increase.

Sub-soiling and ripping can increase crop yields in semi-arid eastern Kenya (AEZ 4 and 5). A yield increase of 117 % of maize total dry matter from 98 to 213 kg/ha in conventional and subsoiled/ripped tillage treatments respectively was recorded in short rains 2007 while 29 % yield increase from 355 to 457 kg/ha of total maize dry matter was recorded in long rains 2009. Use of this technology with integrated soil fertility improvement increased maize grain yield from a mean of 84–108 % because of soil moisture retention and conservation. Fertilizer nutrients applied become available over longer duration of crop growth due to availability of minimum soil water which is required for crop nutrient uptake either by mass flow or diffusion (Tisdale et al. 1985).

The constraints arising from this technology are excess covering of planted seeds with soil. Seeds should be covered lightly depending on size. The cost of sub-soiler and ripper are also prohibitive but this has been solved with a modified mould board plough which is fitted with a shear modification and mould board modification for in situ hardpan breaking and making planting furrow similar to what is made by the subsoiler and ripper. The cost of modification is about US$7.5 and can be fabricated by farmers using locally available materials.

Planting pits (Tumbukiza) resulted in increased maize yields in Mwala district and Mukuyuni division in Machakos district (KASAL) project (2007–2011). A farmer reported a yield of 4–90 kg bags of maize from pits made in 0.25 acres of land in Mukuyuni while another one reported yield of 4 bags of 90 kg maize from 0.25 acres due to pitting in Makutano community association in Katangi Machakos County. Similar observations were made in KASAL project in Mwala district-Mbiuni division, KwaLumbu Village. In the short rains of 2009, a high yielding maize crop was observed planted in planting pits in Kako division of Makueni District. It seemed, pitting increased the yield of the crop as the soils were of low fertility and highly eroded over time. Tumbukiza concentrates rainwater in a smaller area thus raising soil water content per unit volume of soil. This raises the water level in the soil which favors crop nutrient uptake (Tisdale et al. 1985), growth and eventually increased yields. Overall *tumbukiza* can increase crop yields due to rainwater harvesting and conservation.

Deep tillage assists the soil to conserve water for crop growth. Deep tillage increases soil porosity and air spaces which are filled with rain water raising the soil water holding capacity (Hillel 1980). The soil that is not deep tilled would not exhibit such level of porosity and air spaces for rain water holding and hence the difference in sorghum and cowpea yield performance from dry land Katiorin planted during the 1981 short and long rains in those two contrasting managed soils (Table 11.7). This technology is recommended to farmers for increased crop production.

Micro-catchments (runoff-run on) technique involves spreading runoff from part of land on to an adjacent cultivated land without using any structures. The soil in the cultivated area is loosened to increase infiltration. The ratio of the catchment to cultivated area usually varies from 1:1 to 5:1 depending on the rainfall regime, soil properties and crop water requirement. Gibberd (1995), working in semi-arid Eastern Kenya, reported that runoff harvesting using a catchment to cultivated area ratio (C: CA) of 1:1 increased yields of most dry land crops by 30–90 %. Itabari et al. (2004), working with green grams in the same region, reported that a C: CA of 1:2 increased net benefits by 17 % where no furrows were made in the cropped area and by 40 % where connected furrows were made in the cropped area during the long rains of 2002, which had a total of 310 mm of poorly distributed rainfall. The technique has also been shown to substantially increase crop yields in Kitui District (Critchley 1989) and in Baringo County (Table 11.8), where its effectiveness was shown to increase with increasing catchment to cultivated area ratio. In another runoff-run on study in Baringo County, Kinyali et al. (2000) reported that runoff harvesting increased soil water regime by 66 % and the subsequent yield from the runoff treatment was 19 bags per acre whereas the rainfed treatment produced no yield. Manure in combination with semi-circular bunds water harvesting technology resulted in increased grass biomass yields (0.5–3 t/ha) and better ground cover (Munyao et al. 2011). The yields obtained through use of this technology are very high (400 %) yield increase from catchment to cultivated area ratio of 1:1–2:1 (Table 11.8). It is advisable for small scale farmers to use this technology. This technology would however be constrained by lack of land for capturing the rain water. Land next to homestead always has a good crop due to runoff harvesting from the homestead.

Conclusions and Recommendations

There are many technologies for in situ rain water harvesting and their impact is enhanced by combining these technologies with integrated soil fertility improvement. The in situ rainwater harvesting technologies have potential to increase crops and fodder production and it is recommended that, farmers and scientists should be encouraged to test these technologies for wider verification and adoption.

Acknowledgments The authors acknowledge KARI Director for allowing this paper to be presented in the 27th Soil Science Society of East Africa and 6th Africa Soil Science Conference and Centre Director Katumani for giving us the opportunity to compile the information.

References

Critchley WRS (1989) Runoff harvesting for crop production: experience in Kitui District, 1984–1986. In: Thomas DB, Biamah EK, Kilewe AM, Lundgren L, Mochoge BO (eds) Soil and water conservation in Kenya. Proceedings of the third national workshop held at Kabete, Nairobi, Kenya, 16–19 Sept 1986, pp 396–405

Critchley W, Siegert K, Chapman C, Finkel M (1991) A manual for the design and construction of water harvesting schemes for plant production. www.fao.org/docrep/u3160e07.htm

Gibberd V (1995) Final report EMI dry land farming and dry land applied research projects, 1988–1993

Gichangi EM, Njiru EN, Itabari JK, Wambua JM, Maina JN, Karuku A (2007) Assessment of improved soil fertility and water harvesting technologies through community based on-farm trials in the ASALs of Kenya. In: Batiano A (eds) Advances in integrated soil fertility management in Sub-Saharan African: challenges and opportunities. Springer, pp 759–765. doi:10.1007/798-1-4020-5760-1-71

Hillel D (1980) Fundamentals of soil physics. Academic press INC. ISBN 0-12-348560-6, pp 6–17

Imbira J (1989) Runoff harvesting for crop production in semi-arid areas of Baringo. In: Thomas DB, Biamah EK, Kilewe AM, Lundgren L, Mochoge BO (eds) Soil and water conservation in Kenya. Proceedings of the third national workshop held at Kabete, Nairobi, 16–19 Sept 1986, pp 407–431

Itabari JK (1999) Optimizing soil water use in the semi-arid areas of Kenya. In: Efficient soil water use: the key to sustainable crop production in dry areas. Proceedings of the workshops organized by the optimizing soil water use consortium. Niamey, Niger. 26–30 Apr 1998, pp 85–104

Itabari JK, Wamuongo JW (2003) Water-harvesting technologies in Kenya. In: KARI technical note series no. 16

Itabari JK, Kwena K, Esilaba AO, Kathuku AN, Muhammad L, Mangale N, Kathuli P (2011) Land and water management research and development in arid and semi-arid lands of Kenya. In: Bationo A, Waswa B, Okeyo JM, Maina F, Kihara JM (eds) Innovations as key to the green revolution in Africa (2011). Springer, pp 427–438. ISBN 978-90-481-2541-8

Itabari JK, Nguluu SN, Gichangi EM, Karuku AM, Njiru EN, Wambua JM, Maina JN, Gachimbi LN (2004) Managing land and water resources for sustainable crop production in dry areas: a case study of small-scale farms in semi-arid areas of Eastern, Central and Rift Valley Provinces of Kenya. In: Crissman L (eds) Agricultural research and development for sustainable resource management and food security in Kenya. In: Proceedings of end of programme conference, KARI, 11–12 Nov 2003, pp 31–42

Jaetzold R, Schmidt H (1983) Farm management handbook of Kenya. Vol. II/C: natural conditions and farm management information. Ministry of Agriculture/GAT, Nairobi, Kenya

Kathuli P, Itabari JK, Nguluu SN, Gichangi EM (2010) Farmers' perceptions on sub-soiling/ripping technology for rainwater harvesting in mixed dry land farming areas in eastern Kenya. E Afr Agric J 76(2):103–107

Kathuli P, Itabari JK, Nguluu SN, Gichangi EM (2010) Farmers perception on sub-soiling/ripping technology for rainwater harvesting in mixed dry land farming areas in Eastern Kenya, In: Proceedings of the 12th KARI Biennial Scientific Conference, pp 1235–1240, Nairobi, Kenya

Kathuli P, Mweki R (2012) Mbooni west-district (Kinyuani and Syongali cluster sites) farmers capacity building on soil and water conservation. In: KARI annual report 2012, 24–27 Apr 2012

Kilewe AM (1987) Prediction of erosion rates and the effects of top soil thickness on soil productivity. Ph.D. thesis. Nairobi, Kenya. University of Nairobi, p 323

Kinyali SM, Karuku GN, Maina PM, Tirop SK (2000) Rainwater harvesting by runoff-inducement for irrigation of maize production in a semi-arid climate, Baringo district, Kenya. In: Proceedings of the 15th conference of the soil science society of East Africa, held in Nanyuki, Kenya, 19–23 Aug 1996

Kitheka SK, Wambua JRM, Nixon DJ, Itabari JK, Watiki JM, Simiyu SC, Kihumba JN (1995) The use of pitting for the rehabilitation of denuded grazing lands in semi-arid Eastern Kenya. In: Proceedings of the 4th Biennial KARI scientific conference, pp 231–245

Kitinya KT, Onwonga RN, Mbuvi JP, Kironchi G (2011) Tillage and cropping systems: effects on soil moisture and sorghum (*Sorghum bicolour* (L.)) yields in Makueni County, Kenya. In: Paper presented in UoN, College of Agriculture and Veterinary Sciences, faculty of agriculture, -aGRO2011-inaugural biennial conference. Book of abstracts, 26–28 Sept 2011, p 14

Miriti JM, Kironchi JO, Gachene CCK, Esilaba AO, Wakaba PM, Mwangi DM (2009) Effects of water conservation tillage on water use efficiency in maize-cowpea systems in semi-arid Eastern Kenya. E Afr Agric J 74(1):95–101

Muchena FN (1975) Soils of Kampi-ya-Mawe agricultural experimental station. In: Detailed soil survey report no. D1. Nairobi, Kenya: Kenya Soil Survey, p 38

Munyao JN, Kinama JM, Kimenju JW, Ndutu JM, Mutiga ER, Mburu MWK (2011) Rain water harvesting for improved livestock production under climatic variability and change in Kajiado, Kenya. In: Paper presented in UoN, College of Agriculture and Veterinary Sciences, faculty of agriculture, -aGRO2011-inaugural biennial conference. Book of abstracts, 26–28 Sept 2011, pp 19

Mwangi HW, Gachene CCK, Mureith JG, Wamuongo JW, Friesen D (2005) Development and promotion of cover crops in conservation agriculture for increased maize productivity in Kenya. http://www.kari.org

Sombroek WG, Braun HMH, van der Pouw BJA (1982) The exploratory soil map and agro-climatic zone map of Kenya (1980), scale 1: 1,000,000—Kenya Soil Survey E 1, Nairobi

Steiner KG, Rockstrom J (2003) Increasing rain-water productivity with conservation tillage. In: African conservation tillage Network Information series no. 5

Stewart JI, Hash CT (1982) Impact of weather analysis on agricultural production and planning decisions for the semi-arid areas of Kenya. J Appl Met 21:477–494

Tisdale SL, Nelson WL, Beaton JD (1985) Soil fertility and fertilizers. Macmillan Publishing Co, New York, p 102

Chapter 12
Enhancing Food Production in Semi-arid Coastal Lowlands Kenya Through Water Harvesting Technologies

Musyimi B. Muli and Ruth N. Musila

Abstract Two studies were conducted during the long rains of 2010 and 2011 and short rains of 2011 and 2012 to evaluate the performance of drought tolerant maize varieties under different water harvesting technologies (zai pits, tied ridges and conventional). The treatments were laid out in a split plot design with water harvesting methods as the main plots and maize varieties as the sub-plots. Four maize varieties (DK8031, DUMA 43, KDV1 and PH4) were evaluated under the three water harvesting technologies for the first experiment. For the second experiment, four maize plant population treatments of 3, 5, 7 and 9 plants per pit were used. Maize variety DUMA 43 was used alongside the four plant populations. The results for both experiments indicated that the maize yields in zai pits and tied ridges treatments were significantly ($P < 0.05$) higher than conventional treatment and the population of 5 plants per pit had significantly ($P < 0.05$) higher grain yield than the rest of treatments.

Keywords DUMA 43 · Plant population · Tied ridges · Water harvesting · Zai pit

Introduction

The coastal region of Kenya is a food deficit area with households purchasing one third of their food requirements, mainly because of insufficient food production on the farms (Obong'o et al. 1993; Saha et al. 1993). Maize is the most important food stable and constitutes a major component of the diet in the region (Wekesa et al. 2003; Waiyaki et al. 2006). More than 70% of maize area is cultivated by farmers in small holder units of less than 20 hectares of land (Doss et al. 2003). Maize is grown in all agro-ecological zones of the province including arid and semi-arid lands, more suited for sorghum and millet (KARI 2005; Wekesa et al. 2003).

M.B. Muli (✉) · R.N. Musila
Kenya Agricultural Research Institute, Mtwapa P.O. Box 16 MTWAPA, 80109, Kenya
e-mail: musmuli2@gmail.com

W. Leal Filho et al. (eds.), *Adapting African Agriculture to Climate Change*,
Climate Change Management, DOI 10.1007/978-3-319-13000-2_12

Although maize is the most important food crop in the region, the area produces 1.56 million bags against the demand of 3.80 million bags leaving a shortfall of 2.24 million bags (MOA 2012). The deficit has to be imported from other parts of the country or abroad. Over 75 % of the coastal area is either arid or semi-arid (Jaetzold and Schmidt 1983). Rapid population growth in the high and medium potential areas has forced farmers to migrate to marginal areas. Farmers grow local maize varieties recommended for the medium to high rainfall zones for lack of suitable varieties. This has in most cases resulted in crop failures since the drought sets in at the most critical stage of crop growth. Maize grown with zai pits technology showed some yield advantage over that grown with tied ridges and flat planting (Sanginga and Woomer 2009). However, zai pit construction was observed to be laborious and very expensive for farmers without adequate family labour or appropriate equipment (Kabore and Reij 2004). This study was therefore carried out to determine the most adaptable drought tolerant maize variety under various water harvesting methods in semi-arid areas of Kilifi County. The objectives of the study were to determine the most suitable water harvesting technology and determine optimum plant population per zai pit.

Materials and Methods

Site Characterization

The study was conducted at the Kenya agricultural research institute (KARI) testing site of Bamba in Ganze district which is described as a livestock-millet zone in agro-ecological zone (AEZ) coastal lowland (CL) 5 (Jaetzold and Schmidt 1983). Two experiments were carried out during the 2010 long rains, 2011 long rains and, both 2011 and 2012 short rains seasons. The Bamba site is characterized by unreliable and erratic rainfall. The soils are rich in nutrients but crop productivity is limited by high ambient temperatures and low soil moisture. The natural vegetation consists of grass and thickets which is primarily utilized to feed cattle, sheep and goats. Eighty percent of the land is communally owned; communal grazing is practiced while the feeding system for livestock is free range grazing (Ramadhan et al. 2008). The major food crops grown in the area are maize, green gram, cowpea and cassava.

Experiment 1

For experiment 1, four commercial drought tolerant maize varieties were evaluated under three water harvesting methods. The four varieties were DUMA 43, DK8031, KDV1 and Pwani hybrid (PH) 4. PH 4 was used as the local check. The three water harvesting methods included; Zai pits, tied ridges and conventional (flat). Zai pits are holes measuring 60 cm long, 60 cm wide and 60 cm deep. The pits were spaced 60 cm apart within the row and were 90 cm between the rows.

The treatments were laid out in a split plot design with water harvesting methods assigned to main plots and maize varieties assigned to sub-plots. Plots consisted of three rows of zai pits and one row comprised of four pits. At the time of excavation, soil from the first 30 cm was placed separately from that of the next 30 cm (31–60) cm. Before filling up the zai pits, dry grass was laid at the bottom up to about 15 cm and then compacted and the remaining portion of the hole was filled with a mixture of top soil and manure at the ratio of 2:1 up to a depth of 5 cm. Two seeds were planted at 9 equidistant spots within the zai pit and were thinned to one plant per hill resulting in 9 plants per pit.

Ridges were spaced at 90 cm. Crop spacing under ridges and conventional method treatments were 90 × 60 cm. Three seeds were planted per hill and later thinned to two plants per hill. Farm yard manure was applied at the rate of 5 t ha^{-1} during planting. Additional nitrogen fertilizer was applied as a topdress at a rate of 60 kg N ha^{-1} four weeks after planting. To control maize stalk-borer, 'Buddock' (0.05 GR 0.5 g/kg beta cyfluthrin) was applied at the rate of 8 kg ha^{-1} three weeks after planting. The crop in zai pits and tied ridges was weeded once while a second weeding was carried out in the conventional treatment plots.

Data was collected on: stand count, plant height, ear height, ear weight, grain moisture content, stover weight and grain weight. Agronomic data were subjected to analysis of variance and differences among treatment means compared using Fischer's Protected LSD test at $P < 0.05$.

Experiment 2

For experiment 2, four maize plant population treatments of 3 (P3), 5 (P5), 7 (P7) and 9 (P9) plants per pit were used. The density of 9 plants per pit was used as a check. Maize variety DUMA 43 was used alongside the four plant populations. More seeds were planted per hill and were later thinned to required plants per pit. Plots comprised of 3 rows of zai pits and each row comprised of 4 zai pits. Phosphatic basal fertilizer was not applied since farm yard manure had been applied to the zai pits during the 2011 long rains season. Nitrogen fertilizer was applied as a topdress at a rate of 60 kg N ha^{-1} four weeks after planting. To control maize stalk-borer, 'Buddock' (0.05 GR 0.5 g kg^{-1} beta cyfluthrin) was applied at the rate of 8 kg ha^{-1} three weeks after planting. Two weedings were carried out during the crop's growth. Data was collected on: stand count, plant height, ear height, ear length, field weight and moisture content. Maize was also shelled and grain weight recorded.

Results and Discussion

Only the results for the effect of water harvesting technologies on maize performance are going to be presented in this paper. Water harvesting methods significantly influenced all the parameters measured (Table 12.1). Plants in the zai pits treatment were significantly taller than those in tied ridges and the conventional method. This was attributed to enhanced water retention in zai pits compared to the other treatments. However, the plants in zai pits were thinner than those under tied ridges and conventional treatments probably due to plant competition for light in zai pits. There were no significant differences in plant height and ear length between the tied ridges and conventional method. Zai pits and tied ridges treatments produced significantly ($P < 0.05$) higher maize grain yield than the conventional method (Table 12.1). However, there was no significant ($P < 0.05$) difference between the zai pits and the tied ridges. For maize stover, significant ($P < 0.05$) differences were observed among all the water harvesting methods. The zai pit method had significantly higher stover yield than both tied ridges and conventional method. This was also attributed to enhanced water retention in zai pits. The maize stover yield from tied ridges was significantly ($P < 0.05$) higher than from the conventional method. This was because plants in conventional treatments were stunted during the early stages of growth.

Results for experiment two showed differences in plant height but significant ($P < 0.05$) differences existed for only P5 and P7. There was no significant ($P < 0.05$) plant population effect on ear height. However, significant ($P < 0.05$) differences were observed for both ear length and grain yield as indicated in Table 12.2. The populations of 3 plants per pit showed significantly ($P < 0.05$) longer ears than both P7 and the check treatment (P9). However, there was no significant difference between P3 and P5. The longer ear size for P3 was expected because of less competition for nutrients amongst the plants due to lower density than the rest of treatments.

Grain yield ranged from 2.72 to 4.97 t ha^{-1} for the populations of 9 and 5 respectively (Table 12.2). The population of 5 plants per pit showed significantly ($P < 0.05$) higher grain yield than the rest of treatments. The population of 3 plants

Table 12.1 Effect of method of water harvesting on plant height, ear height, grain and stover yield

Water harvesting method	Parameters			
	Plant height (cm)	Ear height (cm)	Grain yield (t ha^{-1})	Stover yield (t ha^{-1})
Zai pits	187.3^{a}	97.7^{a}	2.4^{a}	7.8^{a}
Tied ridges	159.2^{b}	76.3^{b}	2.3^{a}	5.1^{b}
Conventional method	155.2^{b}	74.9^{b}	0.9^{b}	3.5^{c}
LSD	15.9	9.3	0.4	1.5

Means within a column followed by the same superscript are not significantly different ($P \leq 0.05$. The level of significance is 0.05)

Table 12.2 Effect of plant population on plant height, ear height, ear length and grain yield

Plant population	Parameters			
	Plant height (cm)	Ear height (cm)	Ear length (cm)	Grain yield (t ha^{-1})
P7	200.7[a]	85.3[a]	13.3[b]	3.53[b]
P9	195.7[ab]	77.0[a]	13.0[b]	2.72[b]
P3	184.0[ab]	75.6[a]	16.6[a]	3.07[b]
P5	169.6[b]	69.7[a]	14.7[ab]	4.97[a]
LSD	28.47	20.38	2.80	0.94

[a,b,c] Means followed by the same superscript are not significantly different ($P \leq 0.05$) within the column

per pit was too low for the plants to compensate through ear size whereas the population of 9 plants was too high hence presenting high competition for scarce resources. Though we used plant population to reduce competition, other workers have maintained the recommended population of 9 plants per pit while widening the pit from 60 cm × 60 cm to 75 cm × 75 cm (Njiru 2012).

Conclusions

Both zai pits and tied ridges methods of water harvesting are superior to conventional method for maize production and the two technologies would fit well in overcoming the adversaries of climate change in semi-arid lands. The plant population of 5 plants per pit was found to be optimum under the conditions of the current study area. Based on this study, Zai pit technology should be promoted in coastal ASALs with the emphasis of 5 plants per pit instead of 9 plants.

Acknowledgments The authors acknowledge the financial support provided by the Director KARI through both Kenya arid and semi-arid lands and the Kenya agricultural productivity and agribusiness projects. They also thank the Centre Director, KARI-Mtwapa for her support during the study.

References

Doss CR, Mwangi W, Verkuijl H, De Groote H (2003) Adoption of maize and wheat technologies in eastern Africa: a synthesis of the findings of 22 case studies. CIMMYT economics working paper 03–06, Mexico, D. F.: CIMMYT

Jaetzold R, Schmidt H (1983) Farm management handbook of Kenya, vol IIC. East Kenya (Eastern and coast provinces). Farm management branch, Ministry of agriculture, Nairobi

Kabore D, Reij C (2004) The emergence and spreading of an improved traditional soil and water conservation practices in Bukina Faso. Discussion paper for international food policy research institute (IFPRI). 2033 K st.N.W. Washington DC. 20006, 1002 USA

KARI (2005) Research priorities for KARI Mtwapa mandate area (coastal lowland Kenya). KARI Mtwapa

Ministry of agriculture (MoA) (2012) Crops development division. coast province, Annual report, 2012

Njiru JN (2012) Maize production using modified zai pits. Website: http://johnnjue.blogspot.com/2012/05/maize-production-using-modified-zai.html

Obong'o DO, Kangunu FM, Mumu TT, Agesa AA, Jamoza JE, Marwanga W, Grisley W (1993) Farming systems in the Agro-ecological Zones CL2 and CL3 in the Msambweni and Matuga Divisions and CL4 and CL5 in the Kinango Division, Kwale District, Coast Province, Kenya. Ministry of Agriculture, Livestock Development, and Marketing and Kenya Agricultural Research Institute, Kwale District, 1993. Kenya Farming Systems Series No. 4

Ramadhan A, Kiura JN, Njunie MN (2008) Dairy production in coastal Kenya. The current status. Paper presented during the 11 KARI biennial conference at KARI headquarters, 10–14 Nov 2008. Book of abstracts paper no. AP 3. p 10

Saha HM, Kamau GM, Masha JS, Mwangondi Z, Gassare AA, Murage ET, Ndungu GK, Marwanga W, Grisley W (1993) Farming systems in the Agro-ecological Zones CL3 (Malindi Division) CL4 (Ganze Division) CL5 (Kaloleni Division) of Kilifi District, Coast Province. MOALDM and KARI. Kenya Farming Systems Series No. 3

Sanginga N, Woomer PL (eds.) (2009) Integrated soil fertility management in Africa. Principles, practices and development processes. Tropical soil biology and fertility institute of the international centre for tropical agriculture, Nairobi, p 263

Waiyaki N, Mwakubo S, Odhiambo W, Nzuma J, Murithi F, Gamba P, Muhammad L, Nyongesa D, Nambiro E, King'ori B, Obunde P (2006) Determinants of seed maize pricing in Kenya. Kenya institute of public policy research and analysis, Nairobi Kenya. Discussion paper no. 61

Wekesa E, Mwangi W, Verkuijl H, Danda K, Degroote H (2003) Adoption of maize production technologies in the coastal lowlands of Kenya. CIMMYT, Mexico, D.F

Chapter 13
Opportunities for Coping with Climate Change and Variability Through Adoption of Soil and Water Conservation Technologies in Semi-arid Eastern Kenya

L.W. Ngugi, K.P.C. Rao, A. Oyoo and K. Kwena

Abstract Scenario analysis using data generated from APSIM model was conducted to investigate the effect of soil and water conservation practices (tied ridges and mulching) on grain yield of improved maize varieties (Katumani and Makueni) generated with and without N fertilizers under below normal (<250 mm), normal (≥250 < 350 mm) and above normal seasons (≥350 mm) in two sites, Katumani and Makindu in Machakos and Makueni counties Eastern Kenya. Results indicate that the yields were significant (<0.01) under the different seasons and treatments with the magnitude of the yields response varied. Highest yields in Katumani (3,370 kg/ha) were obtained during below normal seasons and when both fertilizer and tied ridges were used. In Makindu, however, under all treatments, highest yields were obtained during above normal seasons with 3,708 kg/ha yield when 40 kg N/ha fertilizer was applied. Lowest yields on the other hand, were obtained during normal seasons in both sites with 507 kg/ha in Katumani and 552 kg/ha under tied ridges and mulching + fertilizers in Makindu. Compared with farmers practice (control), the yield increment obtained was 4 kg/ha (0.6 %) and 5 kg/ha (0.7 %) in Katumani; 32 kg/ha (4.6 %) and 33 kg/ha (4.7 %) in Makindu under mulching and tied ridges respectively during below normal seasons otherwise the yield decreased during normal and above normal seasons with up to 19 % in Makindu when tied ridges was practised. Fertilization increased the yields of maize by as high as 2,552 kg/ha (433 %) and 2,319 kg/ha (166 %) in Katumani and Makindu respectively during above normal seasons. However, during normal seasons, there was yield decrease in Makindu by 42 %.

L.W. Ngugi (✉) · A. Oyoo
International Crops Research Institute for the Semi-arid Tropics, P.O. Box 39063
Nairobi, Kenya
e-mail: lucywangui81@gmail.com

K.P.C. Rao
International Crops Research Institute for the Semi-arid Tropics (ICRISAT),
P.O. Box 5689, Addis Ababa, Ethiopia

K. Kwena
1KARI Katumani, P.O. Box 340-90100, Machakos, Kenya

W. Leal Filho et al. (eds.), *Adapting African Agriculture to Climate Change*,
Climate Change Management, DOI 10.1007/978-3-319-13000-2_13

When both fertilization and soil and water conservation practices was done, yield increase was 2,335 kg/ha (456 %) and 2,382 kg/ha (465 %) in Katumani during normal seasons under mulching +40 kg N/ha and tied ridges +40 kg N/ha respectively. In Makindu, yields declined during normal seasons, however, increase was by 2,229 kg/ha (160 %) and 2,108 kg/ha (152 %) during above normal seasons under mulching +40 kg N/ha and tied ridges +40 kg N/ha respectively. The results indicate that the use of fertilizers and soil and water conservation are indispensable for ensuring food security in semi-arids where rainfall is very variable.

Keywords Tied ridges · Mulching · N fertilizer · Seasons · Katumani · Makindu

Introduction

The agriculture sector is the mainstay of the economies of most of the developing countries, employing about 60 percent of the workforce and contributing an average of 30 % gross domestic product (GDP) in sub-Saharan Africa (World Bank 2011). Kenya in particular, agriculture as the single most important sector in the economy contributed 21.4 and 24 % of the country's GDP in 2010 and 2011 respectively (KPMG Kenya 2012 in Kalungu et al. 2013). Additionally, smallholder farmers provide 75 % of the labour force and 75 % of the market output produce (Alila and Atieno 2006). However, about 82 % of the Kenya's agricultural lands (where 80 % of the smallholder farmers reside and derive their livelihoods, directly or indirectly from agriculture) are marginal with fragile ecosystems receiving variable, irregular and low amounts of rainfall (Mwakubo 2002). The variability in the amount and distribution of rainfall both during and across the seasons, that these areas experience not only makes the availability of water for crop production erratic, but farming, a risky undertaking. Variability in rainfall, the uncertainty that it creates and the general risk averse nature of smallholder farmers acts as a major deterrent to investing on productivity enhancing technologies in the semi-arid regions leading to low yields, worsening food shortages and economic situation of the farmers (Rao and Okwach 2005; Rao 2011). In addition, the low input management practices usually adopted contribute to severe physical, chemical and biological degradation leading to very serious problems of land degradation in the form of soil loss and nutrient depletion posing a threat to food security and sustainability of agricultural production in these areas (Juma 2010). Climate change is expected to further exacerbate this uncertainty and contribute to increased food insecurity in the smallholder farming sector. Small holder farmers, the elderly, women, children, and women and child-headed households will be the most vulnerable because they have limited adaptive capacity (Nyamadzawo et al. 2013). Current projections of future climatic conditions, though not accurate, indicate that there is a high likelihood for an increase in temperature and variability associated with rainfall as well as increased incidence of extreme events (IPCC 2007). Though there are uncertainties over the exact magnitude of these changes, there is a high probability for temperatures to increase by up to 3 °C by end of this century. One of the ways that warmer temperatures are going to manifest is through an increase in crop water requirements due to increased demand for evapotranspiration. Research on how to mitigate these impacts of climate change and variability to agricultural productivity is still very limited. In Kenya, studies have shown that awareness of climate change and variability is still low and farmers have been found to have a problem in differentiating between impacts arising from climate change and problems caused by environmental degradation. This lack of farmer awareness influences negatively on their adoption of appropriate adaptive technologies yet one approach of alleviating the impacts of climate change and variability is through the adoption of appropriate agricultural practices such as soil and water management technologies (Kalungu et al. 2013). These practices are mainly used by farmers with the aim of improving their agricultural production through reducing risks associated with farming. However, some of the agricultural practices continue to reduce the natural protection provided by vegetation cover hence subjecting land to severe soil erosive losses (Khisa et al. 2002). Thus, adopting soil and water management practices influences the agricultural production (Branca et al. 2012). The Government of Kenya is, however, promoting several farming improvement programmes such as the soil and water management project with the aim of increasing soil fertility and crop production (Nyangena 2007). In general, technologies that increase agricultural production and reduce production risk tend to support climate change adaptation as they increase agricultural resilience and reduce yield variability under climate variability and extreme weather events, which might intensify with climate change (Cooper et al. 2008). In Eastern Africa, where annual average precipitation and temperatures are expected to increase with climate change, the greatest impacts on agricultural production are expected from changes in rainfall variability, such as prolonged periods of drought and changes in the seasonal pattern of rainfall (IPCC 2007). Therefore, adaptation strategies that reduce yield variability during extreme events, such as droughts, erratic or changing patterns of rainfall will provide the greatest benefit to farmers and enhance their ability to adapt to current variability and future changes in climate which this study sought to find in semi-arid Eastern Kenya.

Materials and Methods

Study Sites

The study was carried out in two sites in Eastern Kenya; Katumani and Makindu in Machakos and Makueni counties respectively. Katumani falls around KARI Katumani research station at a latitude of 1°34′60S and longitude of 37°15′0E within agro ecological zone (AEZ) IV, which is described as medium to marginal for crop production (Jaetzold and Schmidt 1983). Annual rainfall varies between 500 and 800 mm, with a mean of about 700 mm. The average seasonal rainfall is between 300 and 400 mm for March–May and 310 and 370 mm for October–December while the mean maximum and minimum temperatures are 24.7 °C and 13.7 °C, respectively. However, like other areas of semi-arid eastern Kenya, rainfall in Machakos occurs in events of unpredictable intensity, with coefficients of variation in seasonal rainfall often exceeding 50 % (Keating et al. 1992). Both, the timing and relative lengths of each growing period vary substantially. Makindu lies at latitude of 2°11′38S and a longitude of 37°44′04E within agro ecological zone (AEZ) V. It receives an annual mean rainfall of 614 mm of which 337 mm is received during the short rain season and 195 mm during long rain season. The average annual maximum and minimum temperatures are 28.7 and 17.1 °C respectively. The terrain is flat and 970 m above the mean sea level, the temperature regime at this location is warmer. Both locations experience a semi-arid tropical climate, with a bimodal pattern of rainfall. The long rains come in March to May, with the peak in April followed by an extended dry period which lasts until mid-October when the short rains commence. The short rains peak in November and begin to taper off towards mid-December. They receive low and erratic rainfall with a poor temporal and spatial distribution. Temperature and evapo-transpiration rates are generally high going up to 8.2 mm day^{-1}.

Crop Simulations Using APSIM Model and the Inputs

Long-term yields of maize were simulated using the simulation model APSIM with the weather data available for the two sites, Katumani in Machakos and Kiboko in Makindu. Management practices evaluated were tied ridges, mulching with and without 40 kg N kg/ha of fertilizers under below normal, normal and above normal seasons. In both sites, short rain season (Oct–Dec) was simulated and maize was planted at 4.4 plants per square meter and the sowing window was from mid October to end November each year. A soil profile from Katumani (Chromic Luvisol, Katumani Research Station PAWC = 164 mm) and Kiboko (1,700 mm) was used to simulate soil water and N supply to crops at the two sites respectively. The maize yields simulated were subjected to analysis of variance (ANOVA) and the treatment means separated in linear model using R statistical software.

Results and Discussion

Maize Yields with and Without Fertilizers and Soil and Water Management Practice

The average grain yield of maize obtained with and without fertilizers and soil and water management practices under the different seasons are shown in Table 13.1. The mean grain yield was significantly different under the different treatments, sites

Table 13.1 Simulated mean maize yields (kg) under different fertilizer, soil and water management and seasons at Katumani and Makindu in Eastern Kenya

Site	Katumani			Makindu		
Treatment/Season	<250 (mm)	250–350 (mm)	>350 (mm)	<250 (mm)	250–350 (mm)	>350 (mm)
Farmers' practice	679.4	512.6	588.3	684.2	1,148.2	1,389.4
40 kg N	3,059.6**	2,502.3**	3,140.5**	979.4	656.3	3,708.1**
Mulch	683.5	507.4	562.8	715.8	1,095.5	1,215.7
Tied ridges	683.9	506.9	543.3	717.0	1,093.7	1,114.2
40 kg N-Mulch	3,332.6**	2,848.2**	3,158.9**	1,041.7**	552.5**	3,618.7**
40 kg N-Tied ridges	3,370.0**	2,895.5**	3,123.9**	1,056.4**	527.9**	3,497.2**

Significant at 0.05 '**'

and seasons as shown in Tables 13.2 and 13.3. Among the treatments studied, use of 40 kg N + tied ridges produced the highest maize yields (3,370 kg/ha) followed by 40 kg N/ha + Mulch (3,332.6 kg/ha) during the below normal seasons in Katumani. The lowest grain yields (506 kg/ha) were obtained during normal seasons when tied ridges were used. However, yields obtained with mulching (507.4 kg/ha), farmers' practice (512.6 kg/ha) during normal seasons and tied ridges (543.3 kg/ha), mulching (562.8 kg/ha) and farmers' practice (588.3 kg/ha) during above normal seasons did not differ statistically. Higher maize yields were simulated when soil and water conservation practices along with fertilizer is applied. Significant higher yields are attributed to the higher water harvesting (tied ridges) and retaining capacities giving water more time to penetrate and infiltrate and at the same time nutrients dissolution. Water that could have been lost through runoff and evapotranspiration is utilized by the plants. Makindu on the other hand, highest yields (3,708.1 kg/ha) were obtained during the above normal seasons when 40 kg N was applied followed by (3,618.7 kg/ha) when 40 kg N/ha + Mulch was used. Lowest yields (527.9 kg/ha) were produced when 40 kg N/ha + tied ridges was applied during the normal seasons.

Maize Yields (kg/ha) with Farmers' Practice and Soil and Water Management Practice

In both sites, the yields obtained when farmers' practice and soil and water management practise were applied were not statistically different under the different seasons (Table 13.1). In Katumani, highest yields (683.9 kg/ha) were obtained when tied ridges were applied during below normal seasons. The difference in mean yields obtained when tied ridges, mulch or farmers' practice were applied was very minimal in all the seasons studied. Farmers' practise gave slightly higher yields during normal and above normal seasons compared to mulch and tied ridges. The results suggest that soil and water management does not have beneficial effects on maize during normal or above normal seasons. In Makindu, there is yield incremental trend from below normal seasons to above normal seasons. This indicates the effects of water in this area where rainfall is more variable compared to Katumani. Highest yields (1,389.4 kg/ha) were obtained under farmers' practice. Application of mulch and tied ridges gave the same trend as in Katumani during below normal and normal seasons although the yields were much higher. This implies that whenever fertilizers are not applied, soil and water management may not be worthwhile.

Application of Fertilizers and Use of Soil and Water Management Practices

The mean yields of maize produced with the application of 40 kg N/ha fertilizer and as influenced by soil and water management practices varied from 527.9 to 3,370 kg/ha in both study sites (Table 13.1). The mean yields were significantly different ($P \leq 0.05$) as shown in Table 13.2. In Katumani, interaction of fertilizers and soil and water management practices gave the highest yields during below normal seasons. In Makindu, however, highest yields were obtained during above normal seasons with significantly low yield during normal seasons. The increment in yields 2,382 and 2,229 kg in Katumani and Makindu respectively indicates that integrating fertilizers and soil and water management practices is more effective in achieving higher yields. This was significant as shown in Table 13.3. Additionally,

Table 13.2 Analysis of variance of simulated maize yields (kg) under different fertilizer, soil and water management and seasons at Katumani and Makindu in Eastern Kenya

Source of Error	DF	SS	MS	F value	Pr(>F)
Treatment	5	22,800,265	4,560,053	13.244	$8.97e^{-06}$***
Season	2	5,459,538	2,729,769	7.928	0.00292**
Site	1	1,728,206	1,728,206	5.019	0.03658*
Season*site	2	4,735,601	2,367,800	6.877	0.00533**
Treatment*site	5	6,936,352	1,387,270	4.029	0.01084*

Significance level 0 '***' 0.001 '**' 0.01 '*'

Table 13.3 Analysis of variance of interactions between the treatments, site and seasons under different fertilizer, soil and water management and seasons at Katumani and Makindu in Eastern Kenya

Source of Error	Estimate	Std. Error	*t* value	Pr(>|t|)
40 kg N	2,307.385	479.113	4.816	0.000105***
40 kg N-Mulch	2,519.824	479.113	5.259	$3.80e^{-05}$***
40 kg N-tied ridges	2,536.385	479.113	5.294	$3.51e^{-05}$***
Mulch	−8.873	479.113	−0.019	0.985408
Tied ridges	−15.388	479.113	−0.032	0.974696
250–350 mm	−339.345	338.784	−1.002	0.328475
>350 mm	115.221	338.784	−0.34	0.737326
Makindu	−183.683	553.232	−0.332	0.74333
250–350 mm	319.296	479.113	0.666	0.512751
>350 mm	1,673.364	479.113	3.493	0.002294**
40 kg N	−1,600.08	677.568	−2.362	0.028458*
40 kg N-Mulch	−1,856.16	677.568	−2.739	0.012636*
40 kg N-tied ridges	−1,916.51	677.568	−2.829	0.010380*
Mulch	−56.049	677.568	−0.083	0.934895
Tied ridges	−83.643	677.568	−0.123	0.902986

Significance level 0 '***' 0.001 '**' 0.01 '*'

it should be expected that the benefits obtained from water conservation and fertilization will be much higher in areas and crop seasons with erratic and low total rainfall and with varieties that are more sensitive to soil moisture deficit which is reflected in Katumani. This is because lands in Katumani have been in use for longer period compared to the newly opened lands in Makindu which are more fertile. Highest yield obtained during above normal seasons in Makindu indicates the effectiveness in the use of the fertilizers and conserved water in this area which is slightly drier compared to Katumani.

Conclusion

In general, from the mean maize grain yield data produced both under unfertilized and fertilized conditions of the soil, it could be realized that considerably high yield and monetary benefits would be accrued due to soil and water management practices. This is to be expected from the fact that the water harvested and retained by tied ridges and evapotranspiration minimized through mulching enhances water availability during water deficit periods. In Katumani, results indicate that either the use of fertilizers or soil and water management would be beneficial when done during below and above normal seasons. In Makindu, results have shown that maize requires more available soil water and at the same time tolerates excessive soil water more when grown with fertilizers than without fertilizer applications. Moreover, the substantial yield response of the crop to tied ridging and mulching on both the fertilized and unfertilized simulations indicated that in regions with poor rainfall distributions, soil and water conservation is a necessary agricultural practice in seasons and/or regions receiving variable rainfall.

References

Alila PO, Atieno R (2006) Agricultural policy in Kenya: issues and processes. A paper for the future agricultures consortium workshop, institute of development studies, pp 20–22

Branca B, Tennigkeit T, Mann W, Lipper L (2012) Identifying opportunities for climate smart agriculture investments in Africa. A study for FAO program of work on economics and policy innovations for climate smart agriculture (EPIC)

Cooper PJM, Dimes J, Rao KPC, Shapiro B, Shiferaw B, Twomlow S (2008) Coping better with current climate variability in the rain-fed farming systems of Sub Saharan Africa: an essential step in adapting to future climate change. Agric Ecosyst Environ 126(1–2):24–35

IPCC (2007) Climate change 2007: synthesis report. In: Pachauri RK, Reisinger A (eds) Contribution of working groups I, II and III to the fourth assessment report of the intergovernmental panel on climate change, IPCC, Geneva, Switzerland, p 104

Jaetzold R, Schmidt H (1983) Farm management handbook of Kenya, vol 11c. Natural conditions and farm management information, East Africa, Nairobi; Kenya Ministry of Agriculture and GTZ, Germany

Juma C (2010) The new harvest: agricultural innovation in Africa. Oxford University Press. 296 p

Kalungu JW, Filho WL, Harris D (2013) Smallholder farmers' perceptions of the impacts of climate change and variability on rain-fed agricultural practices in semi-arid and sub-humid regions of Kenya. J Env Earth Sci 3:129–140

Keating BA, Siambi MN, Wafula BM (1992) The impact of climatic variability on cropping research in semi-arid Kenya between 1955 and 1985. In: Probert ME (ed) Sustainable dry land cropping. ACIAR Proceeding, no 41. Canberra, Australia, pp 16–25

Khisa P, Gachene CKK, Karanja NK, Mureithi JG (2002) The effect of post harvest crop cover on soil erosion in a maize-legume based cropping system in Gatanga, Kenya. J Agric Tropics Sub-Tropics 103:17–28

KMPG Kenya (2012) Oil is expected to be a major economic boost in East Africa. Budget brief Kenya, 2012, regional economic overview, pp 1–9

Mwakubo SM (2002). Land tenure and natural resource management in semi-arid areas, Kenya. Agriculture economics department, Egerton University. Paper prepared for the Beijer Research workshop on Property rights, Durban, South Africa, pp 28–30

Nyamadzawo G, Wuta M, Nyamangara J, Gumbo D (2013) Opportunities for optimization of in-field water harvesting to cope with changing climate in semi-arid smallholder farming areas of Zimbabwe. Springer 2(1):100

Nyangena W (2007). Social determinants of soil and water conservation in rural Kenya. Env Dev Sustain 10(6):745–767

Rao KPC (2011). Managing current and future climate induced risk in Eastern and Central African agriculture. Draft paper prepared for the 1st ASARECA general assembly, pp 14–16

Rao KPC, Okwach GE (2005) Enhancing productivity of water under variable climate. Proceedings of the East Africa integrated river Basin management conference, Sokoine University of Agriculture, Morogoro, Tanzania, pp 7–9

World Bank (2011) Global strategy to improve agricultural and rural statistics. Report no. 56719 GBL. World Bank, Washington DC

Chapter 14
Adoption of Water Resource Conservation Under Fluctuating Rainfall Regimes in Ngaciuma/Kinyaritha Watershed, Meru County, Kenya

Evans Mutuma, Ishmail Mahiri, Shadrack Murimi and Peterson Njeru

Abstract A study was carried out to understand the adoption levels of water conservation practices in Meru County. The influence of water resource accessibility on adoption of water conservation (WC) practices and constraints were assessed. Tree planting, roof catchment and bench terraces were the major WC practices in use. Multiple regression analysis revealed that lack of technical know how could explain 83.5 % variations of adoption level of WC practices. One sample t-test comparing the means of WC practices among respondents' was significant at $P < 0.01$. Spearman's rank test revealed a decreasing trend during the long rains (March–May) for the period 1986–2008 at $P < 0.05$. The disparity between the levels of adoption among water users coupled with the decreasing seasonal rainfall calls for urgent and better management of water resources in the study area.

Keywords Rainfall trends · Unsustainable water resource use · Water conservation practices

Introduction

Critical discussions and negotiations on water resources have been on the international agenda and have elevated water resource to a greater global awareness. Water is a scarce resource, yet an essential component for human survival. This scarcity is linked to climate change; demand that exceeds available water resources and most importantly unsustainable use of the resource (Molle 2000). Many parts of the world, markedly the Middle East and the sub Saharan Africa are experiencing intense competition over limited inland water resources. This situation is serious

E. Mutuma (✉) · I. Mahiri · S. Murimi · P. Njeru
Department of Natural Resource Management, Kenya Agricultural Research Institute, P.O Box 14733-00800 KARI-Kabete, Nairobi, Kenya
e-mail: mutumaeva@yahoo.com; mutumaeva1977@gmail.com

W. Leal Filho et al. (eds.), *Adapting African Agriculture to Climate Change*, Climate Change Management, DOI 10.1007/978-3-319-13000-2_14

in shared drainage basins where it has heightened political conflicts (McCartney 2000). The situation in Kenya is not any better. Kenya receives less than 647 m^3 of fresh water per capita per year, making it one of the most water scarce countries in Africa and the world (WRI 1994). Competition over water between agricultural, industrial, domestic and municipal needs has worsened, stretching the recovery of hydrological systems (Orie 1995). Kenya experiences high rainfall variability, low investment in water resources development and poor protection of the existing water resources resulting in extensive degradation (Were et al. 2006).

A basic water management challenge is to find ways to satisfy human needs while coping with climatic changes and protecting the water resource from long-term degradation. With regard to water resource management, the use of participatory approach is one of the principles of the Dublin convention (Cosgrove and Rijsberman 2000). The concept partly reflects the observation that people who inhabit an environment over time are often the ones most able to make decisions about its sustainable use. However, the vast majority of people have become passive observers, and a few people are taking decisions for everyone else. That is one of the prime reasons why the water resources are being destroyed (McLvor 2000). The real revolution in water resources management will therefore come when all stakeholders, where possible, have the power to manage their own water resources. Efforts should be made to maximize productive water use. This could be by finding and stopping wasteful leaks, enhancing focused irrigation techniques, using less water-intensive industrial processes, implementing wastewater recycling, and overall conservation of water catchment areas (Mitchell et al. 2004).

Most of the people living in the rural areas tend to overexploit their immediate environments. According to Gikonyo (2004) indiscriminate cutting of trees has impacted negatively on precipitation and river systems in the Tana Catchment which houses Ngaciuma/Kinyaritha watershed, the study area. Some tributaries in the watershed have become seasonal due to increased land use changes and direct over-abstraction (WRMA 2008). The sub-catchment also experiences temporal variations in water demand that creates a negative balance between demand and supply during the dry season (DAAD 2008).

This study addresses the following key questions: How has been the trend in rainfall for the period 1986–2008 in the study area? What is the level of adoption of water conservation practices in the study area? Which are the constraints faced? Does accessibility and participation in local Water Resource Users Associations (WRUAs) affect adoption of water conservation practices?

Materials and Methods

Study Area

Ngaciuma/Kinyaritha watershed is located within Meru municipality geographically bound by latitudes 37.5°E and 37.75°E and 0.04°N and 0.15°N. The watershed covers an area of 167 km^2 in Imenti North District, Kenya. Climatic conditions range from humid to semi-humid with Agro-ecological zones UM1 = Coffee-Tea Zone, UM2 = Main coffee Zone and UM3 = Marginal coffee zones (Jaetzold et al. 2007). Rainfall is bimodal with mean annual rainfall range of 1,100–1,600 mm and annual temperatures range of 10–30 °C. Altitude ranges from 1,120 to 2,600 m. Geology of the catchment comprises pyrocrasts and the major soils are nitisols which are poorly consolidated hence susceptible to erosion, mass movement and high seepage where water is conveyed in open channels.

Data Collection

Data was negated from primary and secondary sources. Primary sources included administering of questionnaires, focus group discussions, key informant interviews and non-participatory observations. The fieldwork was conducted between the months of June and October 2011. Secondary data included rainfall data acquired from Kenya Meteorological Department in Nairobi and Water Resource Management Authority in Imenti North sub-regional office. Before the main study, a reconnaissance survey was carried out to pre-test the research instruments and work out modalities of identifying respondents in the study area.

Data Analysis

Descriptive statistics were used to analyse socioeconomic parameters. To measure the adoption level of water conservation practices, a weighting system was used that assigned values to each conservation practice based on its importance as perceived by the respondents relative to all other conservation practices. The weighted importance score for each practice was multiplied by reported answers of implementation from respondents. Finally, respondents were categorized as “low”, “fairly low”, “fairly high” and “high” adopters based on the collective ‘adoption score’. Score ranges for low, fairly low, fairly high and high adoption categories were determined by mean and standard deviation, as follows:

Min < A < Mean-St.d: A = Low
Mean-St.d < B < Mean: B = Fairly Low
Mean < C < Mean + St.d: C = Fairly High
Mean + St.d < D < Max: D = High

Trends in rainfall were analyzed using the Ms Excel software to generate graphs. Spearman Rank Correlation test was used to test the null hypothesis of no significant variations in trends of rainfall and stream flow for the period 1986–2008 in

Ngaciuma/Kinyaritha watershed. Rainfall data was ordered and ranked from the lowest to highest. The differences between the rankings were computed and squared. The latter were summed up to yield $\sum \partial i^2$. The Spearman Rank Correlation (rs) was computed using Eq. 14.1.

$$r_s = 1 - 6\sum_{i=1}^{N} \partial i^2 / N(N^2 - 1) \tag{14.1}$$

where, $\partial i = k_i - I, k_i$ is the rank of the series x_i and N the total number of observations. The approximate significance of r_s^2 for $N > 8$ and $df = N - 2$ was calculated by computing:

$$t = rs\ \{df/(1 - r_s^2)\}^{0.5} \tag{14.2}$$

Coefficient of variation was computed to compare variability of each water conservation methods adopted among the respondents. Correlation analysis was used to measure the association between dependent and independent variables. Stepwise multiple linear regression model was used to explain variations in adoption level of WC practices among respondents. Eleven independent variables: age, education level, household size, participation in WRUA conservation activities, farm size, level of information sources and channels, economic motivation, stewardship motivation, level of awareness on sustainable WC practices, attitude towards conservation practices and level of technical knowhow were fitted in model. Backward elimination approach which involved starting with all independent variables and testing them one by one for statistical significance, deleting any that was not significant was used to fit the regression model.

Results and Discussion

Household characteristics for each zone are given in Table 14.1. Majority of the respondents' were farmers at 83.7 %. Due to unreliability of rainfall, rain fed agriculture is no more reliable and thus majority of the farmers are practicing irrigated agriculture. Despite the perception that the study area is well watered, lack of adequate water supplies was the major hindrance to the expansion of irrigation and livestock keeping having up to 75 % of mentions (Table 14.2). This is an indicator that Ngakinya/Kinyaritha watershed is faced with inadequate water supplies against the perception that the watershed is well watered.

Table 14.1 Summary characteristics of selected households according to zones

Household characteristics	Description	Locations			
		Upper zone	Middle zone	Lower zone	Average (%)
Sex (%)	Male	66	82	80	76
	Female	34	18	20	24
Education (%) level of respondents	Primary	50	48.6	54	50.9
	Secondary	26.7	20	25	23.9
	Tertiary	13.3	22.9	10	15.4
	None	10	8.5	11	9.8
Age (%)	18–36	33.3	28	21	27.4
	37–54	50	56.7	66.4	57.7
	>54	17.7	15.3	12.6	15.2
Occupation of household heads	Farmer	90	78	83	83.7
	Civil servants	6.7	12	8	8.9
	Business persons	3.3	10	9	7.4

Table 14.2 Major constraints limiting expansion of irrigation agriculture

Constraint	Frequency of mentions	Percentage of mentions
Land size	60	50
Water accessibility	48	40
Lack of adequate water	90	75
Market problems	6	5
Cost of farm inputs	30	25

Note The percentages do not add to 100 % because the respondents answered to more than one constraint

Rainfall Trends for the Period 1986–2008

Analysis of rainfall for the period 1986–2008 shows an inter-annual fluctuation in rainfall and declining rainfall trends both linearly and in five year moving averages (Fig. 14.1).

The Spearman test showed a significant decreasing trends for the long rains (March–May) for the period 1986–2008 at a statistically significant level of $p < 0.05$ (Table 14.3). This indicates reduction in water resources as rainfall is the major source of water for rivers and ground water replenishment. This would therefore imply inadequate water supply for both domestic and agricultural uses.

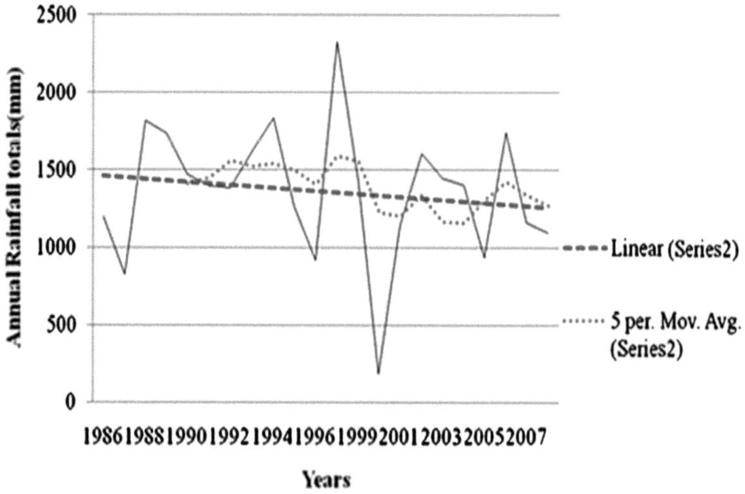

Fig. 14.1 Inter-annual fluctuations in rainfall for the period 1986–2008

Table 14.3 Spearman' test for annual, long (March–May) and short (October–December) rainfall

Variable	Period	1986–2008	1986–1995	1996–2008
	Df	20	8	10
Annual rainfall	r_s	−0.2217	0.2969	−0.2940
	T	−0.9500	0.0857	−0.5620
Long rains(March–May)	r_s	−0.4048	0.0303	−0.2940
	T	−1.9563[a]	0.0857	0.4690
Short rains(October–December)	r_s	0.0350	0.1636	0
	T	0.1566	0.4690	0

[a] Trends statistically significant at $p < 0.05$

Perceived Indicators of Sustainable Water Conservation Practices

Tree-planting; Rainwater harvesting by use of roof catchments; Bench terraces and Mulching are top of the list in terms of prioritization by the respondents (Table 14.4). Drip irrigation according to respondents required a lot of expertise and finances to put up the systems hence the least water conservation method in Ngaciuma/Kinyaritha watershed.

Table 14.4 Prioritized indicators of sustainable water conservation practices

Conservation method	Mean	Standard deviation (Std)	Coefficient of variation (CV)	Priority
Tree planting	0.670	0.140	0.209	1
Rainwater harvesting	0.767	0.161	0.210	2
Bench terracing	0.667	0.300	0.450	3
Mulching	0.500	0.225	0.450	4
Vegetative strips	0.750	0.338	0.451	5
Infiltration ditches	0.308	0.302	0.980	6
Waste water reuse	0.322	0.395	1.227	7
Fanya juu	0.256	0.376	1.469	8
Water metering	0.287	0.422	1.470	9
Drip irrigation	0.156	0.293	1.878	10

Adoption Level of WC Practices and Constraints

Table 14.5 shows the levels of adoption of water conservation practices. It could be inferred from the table that majority of respondents fell into either fairly low or high ranking.

The community faces constraints despite the efforts to participate in water conservation practices (Table 14.6). However, an important finding from this study was that some of the listed adoption constraints decreased with increase in the number of water conservation practices adopted. Similar findings have been observed elsewhere by others (Tenge et al. 2004). Lack of capital was a major constraint. Wealth is linked to power and property rights over natural resources affecting peoples' option for adopting technology (Knox and Meinzen 1999). Those who possess a higher quantity and quality of endowment will place a higher future on medium and longterm benefits produced by investment in water conservation technologies. Majority of the farmers enjoyed security of tenure having title deeds for their land, thus land tenure was not a major constraint.

Table 14.5 Adoption level of water conservation practices by respondents

Group	Scale	Frequency	Percentage of frequency
Group1 (low)	3.817	21	17.5
Group2 (fairly low)	3.818–5.280	48	40.0
Group3 (fairly high)	5.281–6.743	21	17.5
Group4 (high)	7.9	30	25.0
Total		N = 120	100

Max: 7.9, Min: 1.4, Mean: 5.28, Std: 1.463

Table 14.6 Observed constraints in relation to the number of WC practices adopted

		Scores by number of WC practices adopted[a] (%)				
Adoption constraints	Frequency	Overall score (%)	1(n = 8)	2(n = 15)	3(n = 62)	4(n = 35)
Lack of capital	100	83.3	41.65	24.99	8.33	8.33
Lack of technical	72	60.0	28.80	19.8	9.00	2.40
Land tenure insecurity	9	7.5	6.4	1.1	0.00	0.00
Small farm size	90	75.0	22.5	7.5	15.0	30.0
Benefit not known	8	6.7	6.03	0.67	0.00	0.00

[a] Types and numbers of adopted WC measures may have included the following measures either singly or in combination: Fanya juu; bench terraces; grass strips; mulching; tree planting rain water harvesting; waste water reuse (Kitchen gardens) and water metering

Correlation Analysis of Adoption Level of WC Practices and Selective Variables

Table 14.7 shows there is a positive association between adoption of WC practices and information sources as well as communication channels at $P < 0.01$. The positive association implies that as the level of information and communication channels increase, the adoption of WC practices increases and vice versa among farmers' in Ngaciuma/Kinyaritha watershed. Similarly economic motivation/ income level was positively associated $P < 0.01$ with adoption. Sinder and King (1990) in their study on water conservation technologies adoption found that

Table 14.7 Correlation between adoption level of WC practices and selective variables

Variables	Coefficient of variation
Age	−0.033
Education level	−0.013
Household size	−0.013
Participation in WRUA conservation activities	0.126
Farm size	0.353**
Level of information sources and channels	0.460**
Level of economic motivation (Income level)	0.334**
Level of stewardship motivation	0.331*
Level of awareness on sustainable WC practices	0.136
Level of technical know-how	0.918**

* ($P < 0.05$) and ** ($P < 0.01$)

Table 14.8 Regression analysis computing variations in adoption level of WC practices

Description	Label	Water conservation practice	
		B	T
Constant		2.158	7.084**
Level of knowledge (technical know-how)	ə	0.693	18.255*
F = 333.237**	$R^2 = 0.835$	R^2adj = 0.832	

* ($P < 0.05$) and ** ($P < 0.01$)

economic factors promote actual adoption by farmers. There was a positive and significant correlation ($P < 0.01$) between the level of awareness about the effects of water conservation practices and level of adoption. A similar finding was reported by Mahboubi (2005) in their study on factors affecting adoption behaviour of water conservation technologies in Gol watershed in Iran. Similarly the level of knowledge is positively and significantly ($P < 0.01$) correlated with the WC practices adopted.

Regression analysis explaining variations in adoption level of WC Practices In order to explain variations in adoption level of WC practices, stepwise linear regression analysis was used. The results show that the level of technical know-how could explain 83.5 % of variations in adoption level of water conservation activities among respondents (Table 14.8).

According to the results presented in Table 14.8, the following model could be used to explain respondents' adoption level of water conservation practices in the study area:

$$Y = 0.695\partial + 2.157$$

where Y = Dependent variable representing respondents adoption level of water conservation practices and ə is the level of technical know-how of the respondent.

Conclusion and Recommendations

Accelerated water resource degradation is among major constraints to agricultural production. The result of analysis of rainfall indicates reduction in water resources as rainfall is the major source of water for rivers and ground water replenishment. Since adoption of many recommended water conservation measures is still minimal in many areas, paying attention to the factors which determine adoption is a priority. Looking at the factors expected to influence adoption of water conservation practices, participation in Water Resources User Association (WRUA), information sources and channels, economic motivation, level of awareness and level of technical knowhow appear important factors due to their positive and significant correlation with independent variable adoption. These factors interact with each other logically to influence adoption. Additionally the research provides evidence showing that the level of technical knowledge received notable support regarding water conservation practices in the regression model. The findings provided a basis for the following recommendation. It is generally true that access to information sources and communication channels with relevant content may increase awareness about the effects and consequences of water conservation practices among farmers while providing them with required technical knowledge. By understanding the economic and environmental effects of water conservation practices, effective uptake of WC technologies may occur. Thus, provision of required information via various information sources and communication channels in order to raise farmers' awareness is suggested. Community awareness on the conservation measures should be promoted through the use of mass media. The solution is to better target extension services and improve the methods of information delivery. Whereas lack of technical knowhow is cited as a hindrance to adoption, the farmers should be made to know the practices and how best to integrate or incorporate these practices in their agricultural activities for better living as well as protect the environment.

Acknowledgments The authors are grateful to Prof. Chris Shisanya of Kenyatta University who is a partner to EU water facility project that partly funded the project. The KMD and the WRMA sub-regional office in Imenti North District are thanked for availing the rainfall and the respondents' who cooperated during the research survey.

References

Cosgrove WJ, Rijsberman RF (2000) World water vision: making water everybody's business. http://www.worldwatercouncil.org/index.php?id=961. Accessed 7 Dec 2008

DAAD (2008) Water demand in participatory watershed management—Ngaciuma-Kinyaritha watershed, Kenya. DAAD Alumni Summer School Report

Gikonyo J (2004) Modeling of river water abstraction accounting and water allocation in Kenya. National Centre of Competence in Research North-South completed 31 March 2004

Jaetzold R, Hornetz B, Shisanya CA, Schimdt H (2007) Farm management handbook of Kenya, vol III, (Eastern Province), Nairobi

Knox A, Meinzen-Dick R (1999) Property rights, collective action and technologies for natural resource management. CGIAR SP-CAPRi, Washington DC, Policy Brief no. 1, 6p

Mahboubi MR, Irvani H, Rezvanfar A, Kalantari K, Saravi MM (2005) Factors Affecting the Adoption Behaviour Regarding Soil Conservation Technologies in the Zarrin Gol, Watershed in Golestan Province. é Iranian J Nat Resour 57(4):595–606

Mitchell N, Kubota J, Wang G (2004) Effects of land-use changes on hydrological processes in the middle basin of the Heihe River, Northwest China. J Hydrol Process 21:1370–1382

McCartney MP (2000) The water budget of headwater catchment. Phys Chem Earth Part B 25:611–616

McLvor C (2000) Community participation in water management. Experience from Zimbambwe. International and Entwickling GmbH

Molle F (2000) Water pricing in Thailand: Theory and practice. http://www.std.cpc.ku.ac.th.delta/deltacp/waterpricing.html. Accessed on 6 Dec 2008

Orie KK (1995) Legal regime of water allocation and the conservation of aquatic environmental values in Kenya. J East Afr Res Dev 25:76–114

Sinder J, King DA (1990) Adoption of soil and water conservation measures in manilla shire, NSW. Rev Mark Agric Econom 58:2–3
Tenge AJ, De Graaff J, Hella JP (2004) Social and economic factors affecting the adoption of soil and water conservation in West Usambara highlands, Tanzania. Land Degrad Dev 15:99–114
Were E, Jessica R, Swallow B (2006) Local organization and gender in water management: a case from the Kenya Highlands. J Int Dev 20(1):69–81
WRI (1994) World resources: a guide to the global environment. Oxford University Press, New York
WRMA (2008) Report of water abstraction survey in Ngaciuma/Kinyaritha sub-catchment, Kenya by TARDA

Chapter 15
Effects of Integration of Irrigation Water and Mineral Nutrient Management in Seed Potato (*Solanum Tuberosum* L.) Production on Water, Nitrogen and Phosphorus Use Efficiencies

Geofrey K. Gathungu, Joseph N. Aguyoh and Dorcas K. Isutsa

Abstract Inorganic fertilizers have become extremely important in correcting declining soil fertility in seed potato (*Solanum tuberosum* L.) production in Kenya. Unreliable rainfall has also limited seed production. Knowledge on water and nutrient use efficiencies in potato grown under different irrigation regimes with different nitrogen and phosphorus levels will help predict the best application rates for optimal seed potato production and yield. A study was conducted at Egerton University, Horticultural Research and Teaching Farm to determine the effect of integrated application of irrigation water, nitrogen (N) and phosphorus (P) use efficiencies of water, N and P. In a split-split plot design, the irrigation water was applied to maintain soil water at 40, 65 and 100 % field capacity in the main plots, N (0, 75, 112.5 and 150 kg N/ha) to subplots and P (0, 115, 172.5 and 230 kg P_2O_5/ha), which translated into 0, 50.6, 75.9, 101.2 kg P/ha) to sub-subplots, with each treatment replicated three times and the trial repeated once. The irrigation water was applied throughout the potato growth period through drip tube lines, with N supplied as urea (46 % N) in two splits, and P as triple superphosphate (46 % P_2O_5) at planting time. Data on seed potato yield was collected from each treatment at harvest and used to calculate water, N and P use efficiencies. High irrigation water at 100 % compared to 65 and 40 % rate resulted in relatively high N and P use efficiencies, but decreased water use efficiency. Application of intermediate to high N and P nutrient increased the water, N and P use efficiencies. It is recommended to apply low to intermediate irrigation water, intermediate to high N and P to increase their use efficiencies during seed potato production.

Keywords Potato · Nitrogen · Phosphorus · Water use efficiency

G.K. Gathungu (✉) · J.N. Aguyoh · D.K. Isutsa
Department of Crops, Horticulture and Soils, Egerton University, P O Box 536-20115, Egerton, Njoro, Kenya
e-mail: gkgathungu@yahoo.com

G.K. Gathungu · D.K. Isutsa
Department of Plant Sciences, Chuka University, P O Box 109—60400, Chuka, Kenya

W. Leal Filho et al. (eds.), *Adapting African Agriculture to Climate Change*, Climate Change Management, DOI 10.1007/978-3-319-13000-2_15

Introduction

Potato (*Solanum tuberosum* L.) is one of the major food crops and it is the world's fourth important food crop after wheat (*Triticum aestivum* L.), rice (*Oryza sativa* L.), and maize (*Zea mays* L.) because of its high yield potential and nutritive value (FAO 2009; Kumar 2013). It is used to postpone the consumption of cereals and is grown as a security crop against crop failures. It has overtime generated special importance in most parts of Kenya as a means of strengthening food security and increasing revenues for farmers. There is, therefore, increased need to boost output and improve cropping systems to increase profits.

For increased productivity inorganic N fertilizers have become extremely important in correcting declining soil fertility, seed tuber yields and quality. However, use of these fertilizers has not been effective due to isolated application practices. When improperly applied nutrients are not taken up by the crop, use efficiency decreases and residual N and P can be lost through leaching or runoff to groundwater or surface water (Obreza and Sartain 2010). Another factor that has limited seed potato production in many parts of Kenya is unreliable rainfall. Water deficit is a common stress in potato production, which leads to decrease in tuber quality and yield (Hassanpanah 2009). Irrigation has been increasingly employed to curtail effects of drought (Thompson et al. 2007) in many countries, but in Kenya potato farmers rarely use this practice due to cost and lack of knowledge, among other factors. Knowledge on water, mineral nutrient use efficiencies on potato grown under different irrigation regimes, nitrogen and phosphorus will help predict the best application rates for optimal seed potato production and yield. Plant needs for water and nutrients are interdependent, as a good water supply increases nutrient demand and adequate nutrient supply saves water (Roy et al. 2006).

Potato is particularly sensitive to soil water stress (Thompson et al. 2007), which affects physiology, bulking, grade, specific gravity, processing quality and yield of tubers (Shock et al. 2006). A visible shift from rainfed to irrigated seed production could unlock the perennial seed shortage and guarantee food security through increased potato productivity. Therefore proper water and nutrient utilization is a constant concern in improved seed potato production by the informal sector. Farmers in informal seed production sector generally lack knowledge on aspects of water and nutrient management practices that increase their use efficiency as well as productivity and quality of seed potato.

Water-use efficiency (WUE) is a quantitative measure of how much biomass or yield is produced for the amount of water used. WUE is an important determinant of yield under stress and a measure of tolerance to drought and efforts are required to increase production per unit water used, resulting in "more crop per drop" (Blum 2009). Nitrogen uptake efficiency (NUE) is a measure of the capacity of the plant to recover applied N (Errebhi et al. 1999).

However, currently there are no recommendations for the combined use of irrigation water, N and P nutrient application rates for seed potato tuber producers in the informal sector to adopt, leaving the sector to rely on management practices used for commercial potato production. As the need for food production increases with an increasing population growth, it is important that strategies are developed to enhance the nutrient uptake and utilization efficiencies (Liu et al. 2012). Efficient irrigation and nutrient management can maximize marketable seed yield and reduce production costs by conserving water, energy and nitrogen fertilizer. The aim of this study was to establish optimal rates of irrigation water, N and P application for increased production and for improved use efficiency in seed potato tubers production.

Materials and Methods

Potato Growth in the Field

Potatoes were planted in a rain shelter at the Horticultural Research and teaching farm of Egerton University, Njoro between 19th August 2011 and 19th December 2011(Trial I) and repeated between 5th April and 6th August 2012 (Trial II). Three soil samples were randomly collected from the top 0–15 cm and 15–30 cm of the soil profile using a soil auger and analyzed for total N and P before planting and after harvesting of tubers to determine nutrient dynamics. Total nitrogen was determined using the Kjeldahl method (Bremner and Mulvaney 1982). Olsen and Sommers (1982) method was used to determine P content. Soil analysis was conducted at Kenya Agricultural Research Institute (KARI) Njoro Soil Analysis Laboratory. Meteorological data on rainfall, temperature, and relative humidity was obtained from weather stations at Egerton University, Njoro. In the rain shelter, maximum and minimum temperatures were recorded daily from three minimum-maximum thermometers hanged in two extreme ends and the middle of the structure during the trial periods. The minimum temperature was recorded early in the morning and the maximum at midday.

The three factors—water, N and P, were tested in a split-split plot design with irrigation water assigned to main plots, N to subplots and P to sub-subplots. The treatments were replicated three times. The treatments consisted of three irrigation rates (40, 65 and 100 % field capacity), applied throughout the potato growth period through drip tube lines. Water was supplied through irrigating only the root zone, leaving the inter-row spaces dry. A WaterScout (Model SM 100 Sensor) connected to 2,475 Plant Growth Station (Watch Dog Model, Spectrum Technologies, Plainfield, IL 60585, USA) which is applicable between zero percent to saturation was used to indicate the need for irrigation.

Nitrogen was supplied as urea (46 % N) at four rates (0, 75, 112.5 and 150 kg N/ha), each in two splits, with the first half at planting and the second at 5 weeks after planting. Phosphorus was supplied at planting time as triple superphosphate (46 % P_2O_5) at four rates (0, 115, 172.5 and 230 kg P_2O_5/ha), which translated into 0, 50.6, 75.9, 101.2 kg P/ha. Each plot measured 1.8 m × 2.25 m. Each experimental unit consisted of seven rows each with seven tubers. Routine field maintenance practices such as weeding and spraying against diseases and insect pests using appropriate fungicides and insecticides was done when necessary. Seed potato from ten randomly selected plants per treatment were harvested 115 days after planting, labelled and placed in plastic bags for yield determination and consequent calculation of water, N and P use efficiencies.

Water and Nutrient Use Efficiencies

Water use efficiency (WUE), nitrogen use efficiency (NUE) and phosphorus use efficiency (PUE) were calculated as proposed by Tayel et al. (2006) and Roy et al. (2006).

Water Use Efficiency

Water use efficiency = Yield (kg)/water consumptive used (m^3). Consumptive use of water was calculated by summing the amount of water applied per plot during the crop growth period. Total water applied was 18.08, 12.45, and 8.43 m^3 for W1, W2 and W3 irrigation water treatments which after dividing with the 16 sub plots (N × P combinations) translated to 1.13, 0.78 and 0.53 per treatment plot in 100, 65 and 40 % irrigation levels, respectively. This was equal in both Trials I and II. Plot yield was divided by the consumptive water used to obtain the WUE (kg/m^3).

Nitrogen and Phosphorus Use Efficiency

Nitrogen or Phosphorus use efficiency = (Tubers (kg) of fertilized − of controls)/ nitrogen or phosphorus fertilizer used (kg). The amount of nitrogen applied per treatment plot was 0, 75.7, 113.5 and 151.4 grams for N1, N2, N3 and N4 respectively. Phosphorus applied was 0, 20.5, 30.74, and 40.91 grams in treatments P1, P2, P3 and P4 respectively. Before analysis these figures were transformed into kilos. NUE or PUE was expressed in kg/kg.

Data Analysis

Data collected were subjected to analysis of variance using the SAS system for windows V8 1999–2001 by SAS Institute Inc., Cary, NC, USA and significantly different means separated using Tukey's Studentized Range Test at $P \leq 0.05$.

The data on nitrogen and phosphorus use efficiencies were first transformed using the square-root transformation before analysis using the formula: transformed data = square root (collected data + 1) to enable normal distribution and the homogeneity of variances before analysis.

Results

Soil Analysis and Climatic Data

The soil had a pH of 5.46, total N of 0.12 %, available P of 0.19 %, exchangeable K of 0.10 % and organic carbon of 3.51 % in the 0–15 cm layer and a pH of 5.6, total N of 0.02 %, available P of 0.11 %, exchangeable K of 0.08 % and organic carbon of 3.02 % in the 15–30 cm layer. A total of 601.6 mm and 942.3 mm of rain was received in the site during the first (August–December 2011) and the second (April–August 2012) Trial respectively. A total of 635 and 221.7 mm was received in the site and mean temperatures were 19.0 and 18.9 °C during Trials I (April–June 2012) and II (November 2012–January 2013), respectively (Table 15.1). In the rain shelter the mean temperatures were 20.7 and 20.5 °C in Trials I and II, respectively (Table 15.2).

Water Use Efficiency

Water use efficiency (WUE) significantly related to all tested factors. Integrating N with either irrigation or P significantly affected WUE. High irrigation water alone reduced WUE. The WUE was 10.81 and 12.67 kg/m^3 when moisture was maintained at 100 % FC compared to 16.91 and 20.75 kg/m^3 obtained with low irrigation water in Trials I and II, respectively. Low irrigation at 40 % followed by 65 % had the highest and 100 % had the least WUE. High irrigation water at 100 % compared to 40 % reduced WUE by 6.1 and 8.08 kg/m^3 (Fig. 15.1). Application of N irrespective of irrigation increased WUE at all levels in both Trials. Higher WUE was recorded when irrigation was combined with application of 150 kg N/ha. The lowest WUE was recorded when 100 % irrigation was integrated with 0 kg N/ha in both Trials. Contrary to irrigation, application of both N and P significantly increased WUE. N and P rates increased WUE from 8.48 and 9.84 kg/m^3 to 14.13 and 16.20 kg/m^3 in Trials I and II, respectively (Table 15.3).

Nitrogen application irrespective of P rate improved WUE. Increasing N from 0 to 150 kg N/ha increased WUE from 11.1 and 12.88 kg/m^3 to 18.04 and 20.73 kg/m^3 in Trials I and II, respectively. Similarly, application of P from 0 to 101.2 kg P/ha increased WUE irrespective of N rate from 11.67 and 14.04 kg/m^3 to 17.42 and 20.5 kg/m^3 in Trials I and II, respectively. WUE was greatly reduced by low N and P rates (Table 15.3).

Table 15.1 Climatic data from 2011 to January 2013 at Egerton University meteorological station (9035092)

Month	Year								
	2011			2012			2013		
	R (mm)	T (°C)	RH (%)	R (mm)	T (°C)	RH (%)	R (mm)	T (°C)	RH (%)
January	3.3	21.2	53	0	21.1	40	42.7	20.6	53
February	9.6	22.3	42	16.3	21.3	45	2.5	21.9	42
March	182.3	21.4	53	31.6	22.5	42	85.4	21.3	52
April	20.9	21.0	53	287.0	20.0	70			
May	116	20.5	66	181.8	19.7	71			
June	216.5	19.3	74	166.2	18.7	74			
July	130.1	19.1	74	87.2	17.6	78			
August	130	18.2	74	220.3	18.7	69			
September	149.3	18.6	70	192.4	19.4	65			
October	89.2	19.8	65	94.3	20.0	62			
November	146.7	19.0	75	26.6	19.7	66			
December	86.4	19.3	61	152.1	19.3	65			
Total	1280.3			1455.8					

R rainfall, *T* temperature, *RH* relative humidity

Table 15.2 Mean monthly temperature data in the rain shelter for 2011 (August–December) and 2012 (April–August) seasons

Month	2011			2012		
	Maximum (°C)	Minimum (°C)	Mean (°C)	Maximum (°C)	Minimum (°C)	Mean (°C)
April	–	–	–	23.1	16.2	19.7
May	–	–	–	23.6	16.9	20.3
June	–	–	–	24.2	17.4	20.8
July	–	–	–	23.7	18	20.9
August	22.6	16.2	21.4	23.6	17.9	20.8
September	23.1	16.9	21.1	–	–	–
October	24.9	16.5	20.7	–	–	–
November	24.1	16.3	20.2	–	–	–
December	23.9	16.1	20.0	–	–	–
Mean	23.7	16.4	20.7	23.6	17.3	20.5

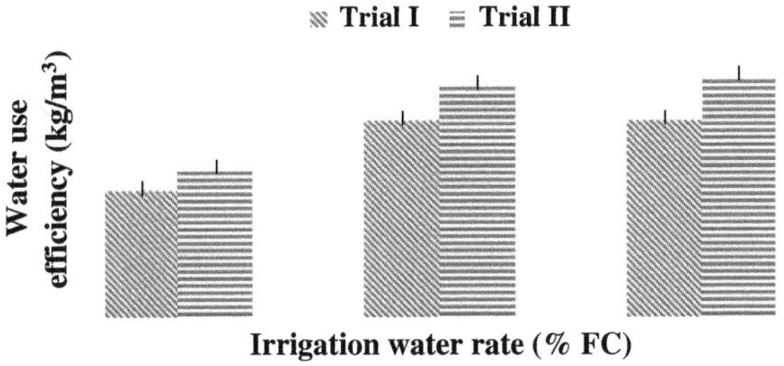

Fig. 15.1 Effect of irrigation water on potato water use efficiency

Nitrogen Use Efficiency

Irrigation water, N and P rates significantly affected NUE. Integrating N or P with irrigation water significantly affected NUE. Application of irrigation water where low N rate of 0 kg N/ha was supplied led to zero NUE. NUE increased with irrigation water and N rates. High irrigation water application resulted in better NUE compared to low irrigation water. Increasing irrigation rate from 40 to 100 % increased NUE by 14.42 and 13.3 kg/kg in Trials I and II, respectively (Table 15.4).

Irrespective of irrigation rate, N increased NUE up to 112.5 kg N/ha beyond which it declined. Generally application of N beyond 112.5–150 kg N/ha decreased NUE by 1.95 and 14.48 kg/kg in Trials I and II, respectively. Application of N and P also improved the NUE regardless of the P rate. Like irrigation, P alone reduced NUE to zero. Integration of high compared to low N and P rates improved NUE. Increasing P rate from 0 to 101.2 kg P/ha regardless of N rate increased NUE from 21.67 and 28.17 kg/kg to 30.45 and 42.69 kg/kg, which was equivalent to 8.78 and

Table 15.3 Effect of N and P rates on potato water use efficiency

N rate (kg N/ha)	P rate (kg P/ha)					Irrigation water (% FC)			
	0	50.6	75.9	101.2	Mean	100	65	40	Mean
Trial I									
0	8.48d*	10.25d	12.26d	13.06d	11.01	7.74d	12.18d	13.12d	11.01
75	11.14c	12.99c	15.41c	15.72c	13.82	10.10c	15.77c	15.58c	13.82
112.5	12.94b	15.45b	18.45b	19.48b	16.58	12.25b	18.52b	18.97b	16.58
150	14.13a	16.75a	19.87a	21.43a	18.04	13.15a	21.03a	19.95a	18.04
Mean	11.67	13.86	16.50	17.42		10.81	16.87	16.91	
MSD (N)	0.78								
MSD (P)	0.78								
MSD (W)	0.61								
CV (%)	8.42								
Trial II									
0	9.84c	12.44c	15.12d	14.13c	12.88	8.90c	14.25c	15.50c	12.88
75	13.31b	16.21b	18.65c	18.64b	16.70	11.98b	18.59b	19.54b	16.70
112.5	16.83a	18.80a	22.77b	24.98a	20.84	14.97a	23.64a	23.92a	20.84
150	16.20a	18.96a	23.49a	24.27a	20.73	14.82a	23.33a	24.05a	20.73
Mean	14.04	16.60	20.01	20.50		12.67	19.95	20.75	
MSD (N)	0.76								
MSD (P)	0.76								
MSD (W)	0.60								
CV (%)	6.90								

MSD minimum significant difference

*Means followed by the same letter(s) along the column for different P or irrigation rates by N rates are not significantly different at $P \leq 0.05$ according to Tukey's studentized range test

Table 15.4 Effect of irrigation water, N and P rates on potato NUE

N rate (kgN/ha)	Irrigation water (% FC)				P rate (kg P/ha)				
	100	65	40	Mean	0	50.6	75.9	101.2	Mean
Trial I									
0	0.00c*	0.00c	0.00c	0.00	0.00b	0.00c	0.00c	0.00c	0.00
75	35.12b	36.99b	17.26b	29.80	27.95a	28.77b	32.06b	30.41b	29.80
112.5	44.82a	43.55a	27.34a	38.58	30.15a	35.63a	41.89a	46.60a	38.57
150	40.38a	45.62a	23.89a	36.63	28.56a	33.70a	39.45a	44.79a	36.63
Mean	31.54	30.08	17.12		21.67	24.53	28.35	30.45	
MSD (N)	4.76								
MSD (P)	4.76								
MSD (W)	3.75								
CV (%)	15.21								
Trial II									
0	0.00c	0.00c	0.00c	0.00	0.00c	0.00c	0.00c	0.00d	0.00
75	46.08b	44.72b	28.27b	39.69	35.67b	37.32b	39.49b	46.28c	39.69
112.5	60.44a	64.57a	39.33a	54.78	44.65a	48.35a	53.31a	72.80a	54.78
150	44.19b	46.80b	29.91b	40.30	32.36b	33.88b	43.30b	51.66b	40.30
Mean	37.68	39.03	24.38		28.17	29.89	34.03	42.69	
MSD (N)	5.42								
MSD (P)	5.42								
MSD (W)	4.27								
CV (%)	14.52								

MSD minimum significant difference

*Means followed by the same letter(s) along the column for different irrigation water and P are not significantly different at $P \leq 0.05$ according to Tukey's studentized range test

Table 15.5 Effect of irrigation water, N and P rates on potato PUE

N rate (kg/ha)	Irrigation water (% FC)				P rate (kg P/ha)				
	100	65	40	Mean	0	50.6	75.9	101.2	Mean
Trial I									
0	71.6b*	66.27b	48.99a	62.28	0.00a	66.78b	97.16c	85.17c	62.28
75	77.47b	72.34b	50.31a	66.71	0.00a	69.81b	107.28b	89.74b	66.71
112.5	111.46a	105.8a	59.23a	92.15	0.00a	97.13a	140.68a	130.8a	92.15
150	122.29a	123.0a	55.3a	100.19	0.00a	104.7a	150.80a	145.3a	100.19
Mean	95.70	91.85	53.45		0.00	84.61	123.98	112.74	
MSD (N)	17.54								
MSD (P)	17.54								
MSD (W)	13.82								
CV (%)	20.04								
Trial II									
0	89.50bc	80.70b	63.78a	77.99	0.00a	95.31a	132.78c	83.88d	77.99
75	79.39c	94.05b	75.97a	83.14	0.00a	103.35a	128.7d	100.5c	83.14
112.5	93.12b	124.3a	79.78a	99.05	0.00a	74.82b	151.1b	170.3a	99.05
150	133.25a	123.3a	65.84a	107.47	0.00a	106.54a	186.63a	136.7b	107.47
Mean	105.58	98.31	71.34		0.00	95.01	149.81	122.83	
MSD (N)	15.96								
MSD (P)	15.96								
MSD W)	12.58								
CV (%)	14.84								

MSD minimum significant difference

*Means followed by the same letter(s) along the column for different irrigation water and P rates are not significantly different at $P \leq 0.05$ according to Tukey's studentized range test

14.52 kg/kg in Trials I and II, respectively. The highest NUE was recorded when 112.5 kg N/ha was integrated with 101.2 kg P/ha. Regardless of P rate, high N rate of 150 kg N/ha reduced NUE in both Trials (Table 15.4).

Phosphorus Use Efficiency

Irrigation, N and P rates significantly affected the phosphorus use efficiency (PUE). Furthermore, integrating N or P with irrigation water significantly affected PUE (Table 15.5). Application of irrigation water, N and P increased PUE. Increasing irrigation water rate from 40 to 100 % regardless of N rate significantly increased PUE from 53.45 and 71.34 kg/kg to 95.70 and 105.58 kg/kg, which was equivalent to 42.25 and 34.24 kg/kg in Trials I and II, respectively (Table 15.5).

Increasing N from 0 to 150 kg N/ha increased PUE from 62.28 and 77.99 kg/kg to 100.19 and 107.47 kg/kg, which was equivalent to 37.91 and 29.48 kg/kg in Trials I and II, respectively. Integration of irrigation and N rates increased PUE. However, PUE decreased when N rate was increased to 150 kg N/ha and integrated with lowest irrigation rate of 40 %. Application of P increased PUE and regardless of N rate, low P at 0 kg P/ha led to zero PUE. PUE increased with P rate from 0 to 75.9 kg P/ha beyond which additional P to 101.2 kg P/ha decreased PUE by 11.24 and 26.98 in Trials I and II, respectively (Table 15.5).

Discussion and Conclusions

The potato NUE and PUE increased, while WUE decreased with irrigation water, N and P rates. Elsewhere, WUE has been reported to decrease with the increase of amount or frequency of irrigation (Amanullah et al. 2010; Badr et al. 2012). In this study, application of 100 % irrigation water alone compared to 40 % irrigation water did not lead to high WUE. This probably is due to the nutrient limitations leading to reduced crop growth and available N and P uptake.

Nitrogen and phosphorus application improved WUE. Badr et al. (2012) reported improved WUE with N supply in potato, but decreased WUE as the irrigation rate was increased. This shows that supply of N and P in potato cropping systems is essential under water non limiting conditions as evidenced by high WUE observed under high N and P conditions. This suggests that where low N and P rates were supplied, potato plants did not fully utilize available soil moisture and consequently growth and development were reduced, resulting in low yields. Sufficient quantities of P have been reported to stimulate early root growth and WUE. The low WUE with high irrigation alone and high WUE with high N and P rates probably indicate that better plant performance requires supply of water, N and P at optimal levels. Additionally, nutrient imbalances influence uptake of nutrients even when available abundantly. Vitousek et al. (2009) concluded that nutrient additions to intensive agricultural systems range from inadequate to excess and that nutrient imbalance is a serious problem in soils. Irrigation regime is crucial in determining plant ability to take up the N available in the soil since a well-watered crop is more capable to take advantage of the applied fertilizer (Costa et al. 1997). This aspect is particularly relevant for estimating WUE at different irrigation, N and P rates and consequently their impact in seed potato production.

The 0 kg N/ha regardless of the irrigation water and P rates lead to zero NUE. This was similar to 0 kg P/ha, which also led to zero PUE. These results suggest that no supply of either N or P does not improve their use, because if their levels are limiting within the soil, this will reflect on the final seed potato yield. High compared to low irrigation rates led to high NUE or PUE. Liu et al. (2012) indicated that soil water and fertilizer management are important in enhancing N uptake and utilization efficiency through reduction of losses in ammonia volatilization. Probably, where high irrigation water was applied, there was more of the soil water available, which resulted in better uptake and utilization of the N and P applied and consequently increasing their use efficiency by the potato plant in growth and development. Balancing irrigation water, N and P rates is one of the key factors that influence N and P uptake and use. Optimal irrigation water application could significantly reduce any possible loss of both N and P and thus enhance their use. This result suggests that synchronized application of irrigation water, N and P is advantageous both in improving their availability and utilization, as well as seed potato tuber production and quality. Water deficit in soil may affect nutrient availability and absorption by plant roots (Roosta et al. 2009). It is, therefore, crucial to understand that combining water and nutrient use efficiencies improves growth, yield and quality of seed potato tubers.

It is concluded that high irrigation water supply resulted in relatively high N and P use efficiencies, but decreased water use efficiency. However, application of intermediate to high N and P nutrient improved the water, N and P use efficiencies. Therefore it is recommended to apply up to intermediate irrigation water but high N and P to increase their use efficiencies in seed potato production.

Acknowledgments The authors acknowledge the National Commission for Science, Technology and Innovation for funding this study and Egerton University for logistical support.

References

Amanullah ASM, Talukder SU, Sarkar AA, Ahsanullah ASM (2010) Yield and water use efficiency of four potato varieties under different irrigation regimes. Bangladesh Res Publ J 4:254–264

Badr MA, El-Tohamy WA, Zaghloul AM (2012) Yield and water use efficiency of potato grown under different irrigation and nitrogen levels in an arid region. Agric Water Mgt 110:9–15

Blum A (2009) Effective use of water (EUW) and not water-use efficiency (WUE) is the target of crop yield improvement under drought stress. Field Crops Res 112:119–123

Bremner JM, Mulvaney CS (1982) Nitrogen-total. In: Page AL, Miller RH, Keeney DR (eds) Methods of soil analysis, part 2. Chemical and microbiological properties, 2nd edn. Number 9 (part 2), Agronomy, Am. Soc. of Agron. Inc, Soil Sci. Soc. of Am, Inc. Madison, Wisconsin, USA, pp 595–622

Costa LD, Vedove GD, Gianquintoi G, Giovanardi R, Peressotti A (1997) Yield, water use efficiency and nitrogen uptake in potato: influence of drought stress. Potato Res 40:19–34

Errebhi M, Rosen CJ, Lauer FI, Martin MW, Bamberg JB (1999) Evaluation of tuber-bearing Solanum species for nitrogen use efficiency and biomass partitioning. Am J Potato Res 76:143–151

FAO (2009) New light on a hidden treasure: international year of the potato 2008. An end of year review, 144 pp

Hassanpanah D (2009) In vitro and in vivo screening of potato cultivars against water stress by polyethylene glycol and potassium humate. Biotechnology 8:132–137

Kumar CV, Prakash SS, Prashantha GM, Kumar MBM, Lohith S, Chikkaramappa T (2013) Dry matter production and yield of potato as influenced by different sources and time of fertilizer application and soil chemical properties under rainfed conditions. Res J Agric Sci 4:155–159

Liu G, Li Y, Alva AK, Porterfield DM, Dunlop J (2012) Enhancing nitrogen use efficiency of potato and cereal crops by optimizing temperature, moisture, balanced nutrients and oxygen bioavailability. J Plant Nutri 35:428–441

Obreza TA, Sartain JB (2010) Improving nitrogen and phosphorus fertilizer use efficiency for Florida's horticultural crops. HortTechnol 20:23–33

Olsen SR, Sommers LE (1982) Phosphorus. In: Page AL, Miller RH, Keeney DR (eds) Methods of soil analysis, part 2. Chemical and microbiological properties, 2nd edn. Number 9 (part 2), Agronomy, Am. Soc. of Agron. Inc, Soil Sci. Soc. Of Am, Inc. Madison, Wisconsin, USA, pp 403–427

Roosta HR, Shahnazari A, Nazari F (2009) Comparative effects of conventional irrigation (CI) and partial root zone drying (PRD), and various sources of nitrogen on growth and yield in potato under field condition. Am-Eurasian J Sustain Agric 3(4):643–651

Roy RN, Finck A, Blair GJ, Tandon HLS (2006) Plant nutrition for food security: a guide for integrated nutrient management. In: Food and agriculture organisation of the united nations, fertiliser and nutrition bulletin. Rome, Italy, 348 pp

Shock C, Flock R, Eldredge E, Pereira A, Jensen L (2006) Successful potato irrigation scheduling: sustainable agriculture techniques. Oregon State University Extension Service. EM8911E, October 2006

Tayel M, El-dardiry E, El-Hady MA (2006) Water and fertilizer use efficiency as affected by irrigation methods. Am-Eurasian J Agric Environ Sci 1(3):294–301

Thompson AJ, King JA, Smith KA, Don HT (2007) Opportunities for reducing water use in agriculture. Defra research project final report for WU0101

Vitousek PM, Naylor R, Crews T, David MB, Drinkwater LE, Holland E, Johnes PJ, Katzenberger J, Martinelli LA, Matson PA, Nziguheba G, Ojima D, Palm CA, Robertson GP, Sanchez PA, Townsend AR, Zhang FS (2009) Nutrient imbalances in agricultural development. Science 324:1519–1520

Chapter 16
Integrating Farmers and Scientific Methods for Evaluating Climate Change Adaptation Options in Embu County

P.N.M. Njeru, J. Mugwe, I. Maina, M. Mucheru-Muna, D. Mugendi, J.K. Lekasi, S.K. Kimani, J. Miriti, V.O. Oeba, A.O. Esilaba, E. Mutuma, K.P.C. Rao and F. Muriithi

Abstract Potential for promoting sorghum crop as a climate change adaptation strategy for rain-fed agriculture in Embu County, Kenya was evaluated using farmer perceptions and scientific methods. Three hundred and sixty six smallholder farmers participated in the evaluation. The treatments which were overall rated as 'good' are tied ridges with a mean score of 2.9 and mean rank (2,873.87). Under this treatment sorghum grain yield of 3.7 t ha^{-1} was recorded with application of 40 kg P ha^{-1} + 20 kg N ha^{-1} + Manure 2.5 t ha^{-1}. This was closely followed by tied ridges and contour furrows overall rated 'good' best three under the same soil fertility management options with a mean score ranging from 2.65 to 2.8 and yielding 2.7–3.7 t ha^{-1}. However, the treatments which were rated as 'poor' were experiment controls with a mean score below (1.43), mean rank (1,101.24) and yielding as low as (0.7 t ha^{-1}). Therefore, integration of organic and inorganic inputs under various water harvesting technologies could be considered as an alternative option towards food security under climate change for semi-arid areas of Embu County.

Keywords Climate change mitigation · Food security · Rain-fed agriculture · Soil amendments

Introduction

Agricultural productivity has been impaired by climate change, declining soil fertility, degradation of natural resources, inefficient markets, weak institutions and policies in semi-arid areas of Kenya. In Kenya, over 13 million of the 38 million

P.N.M. Njeru (✉) · J. Mugwe · I. Maina · M. Mucheru-Muna · D. Mugendi · J.K. Lekasi · S.K. Kimani· J. Miriti · V.O. Oeba · A.O. Esilaba · E. Mutuma · K.P.C. Rao · F. Muriithi
Department of Natural Resource Management, Kenya Agricultural Research Institute, KARI Muguga South, P.O. Box 30148-00100 Nairobi, Kenya
e-mail: njerupeterson@yahoo.com; njerupeterson1@gmail.com

W. Leal Filho et al. (eds.), *Adapting African Agriculture to Climate Change*, Climate Change Management, DOI 10.1007/978-3-319-13000-2_16

people live below the poverty line of less than U.S$1 a day. Agriculture is the mainstay of the Kenyan economy contributing approximately 55 % of Gross Domestic Production (GDP). The sector further provides 80 % employment, accounting for 60 % of the exports and 45 % of the government revenue (Ragwa et al. 1998). The government in Kenya has put in place the Agricultural Input Subsidy Program (AISP) to support farmers so that they can access inputs such as inorganic fertilizers. In its "Vision 2030", the government also spells out the desire to use agriculture as the vehicle to transform the country to industrialization (CAADP 2008). However, more than 80 % of Kenya is classified as arid and semi-arid areas characterized by low and erratic waterfall, high evaporation rates and soils that are unsuitable for sustainable rain-fed agriculture (Miriti et al. 2012; Fongod et al. 2012).

Failure of rainfall is the main cause of persistent rural poverty (Miriti et al. 2012). The dry spell analysis indicates that potentially yield-limiting dry spells occur at least in 75 % of the seasons during a 20-year period (GoK 2007). Drought is also another risk to crop failure which has led to reluctance by farmers to invest on crop land (KARI 2009). Therefore irrigation, to maintain soil water content within the plant root zone at an optimal level may be the only option for adapting to climate change in these areas. However, the same is not feasible to most small-holder farmers because they either lack resources to invest in irrigation technologies or water not available for irrigation. This situation could be ameliorated through adoption of on-farm rain water harvesting and integrated nutrient management techniques as alternative option for mitigating impacts of prolonged dry spells. Incorporating highly valued traditional crops that are tolerant to drought into these farming systems is another option.

The challenge now remains on how to maximize agricultural production in semi-arid areas of Embu County in the face of climate change. The low crop production is also often associated with lack of appropriate farming practices that are suited to the fragile ecosystems to cope with climate change challenges (Bationo et al. 2004; Mbogoh 2000). Most of the smallholder farms are characterized by nutrient mining as a result of harvest and residue removal (Mugendi et al. 2003; Bielders et al. 2002) as well as lack of resources to invest in mineral fertilizers. Very little nutrient replenishment is practiced in Eastern Kenya (Mugendi et al. 2010). The recommendation of African Fertilizer Summit (2006) 'to increase the fertilizer use from the current 8–50 kg ha^{-1} nutrient by 2015' reinforces the role of fertilizer as a key entry point for increased crop productivity and attaining food security in Embu County. However, most farmers cannot afford to buy inorganic fertilizers due to their high prices (Sanginga et al. 2009; Crews and Peoples 2004). These inappropriate farming systems practiced by farmers are leading to land degradation and lack of appropriate rain water harvesting and conservation technologies are resulting in low crop yields (Njeru et al. 2011a, b; Kimani et al. 2007). Therefore, food security situation is expected to continue deteriorating and could worsen in future if climate change adaptation options are not taken up quickly in semi-arid areas of Embu County. Therefore this study assessed the comparison of farmer's evaluation and scientific method on various climate change mitigation technologies for increased sorghum productivity at Kiritiri Division, Mbeere District in Embu County.

Materials and Methods

Study Site Location and Description

The study was conducted in Kiritiri Division, Mbeere South Sub-County which lies in the southeastern slopes of Mt. Kenya (Fig. 16.1). It lies between latitude 0.91672°S and Longitude 37.4768°E to the North and between Latitude 0.4733°S and Longitude 37.91238°E to the South. The district lies at an altitude of 800 m a.s. l with an average rainfall of 700–900 mm, temp of 21.7–22.5 °C. The predominant soil type is ferralsols. Kiritiri division, Mbeere south District is generally a low potential dry zone. It is covered by three agro-ecological zones; the marginal cotton zone (LM4); the lower midland livestock-millet zone (LM5); and the lowland livestock millet zone (L5) (Jaetzold et al. 2007). The study was conducted in agro-ecological zone (LM4/5) in Long rains 2011, 2012 and short rain 2011.

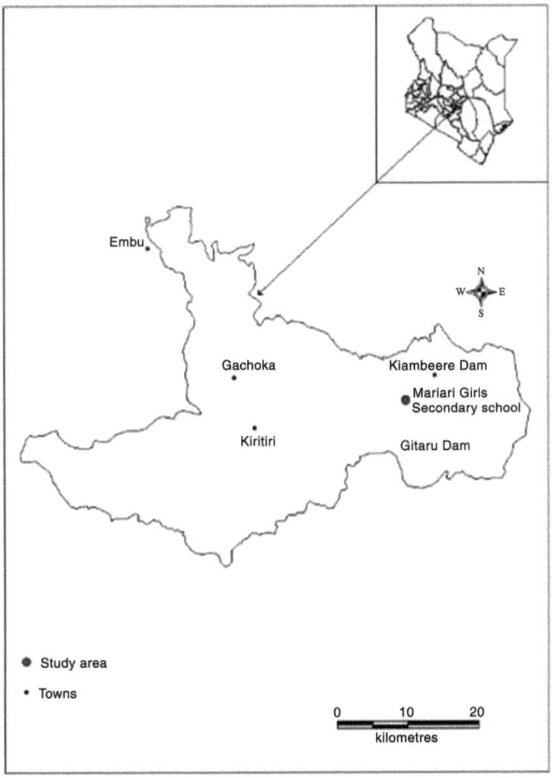

Fig. 16.1 Shows the study site in Mbeere South Sub-County in the map of Kenya

Experimental Design

The treatments were arranged in a factorial structure, each treatment being a combination of one of the 3 levels of water harvesting techniques (Tied Ridges, contour furrows and conventional tillage/farmers Practice), 2 levels of cropping systems (Sole sorghum-Gadam, Sorghum and cowpea (M66) intercrop) and 6 levels of soil fertility amendment options (Control, 40 kg P ha^{-1} + 40 kg N ha^{-1}, 40 kg P ha^{-1} + 20 kg N ha^{-1}, 40 kg P ha^{-1} + 40 kg N ha^{-1} + Manure 5 t ha^{-1}, 40 kg P ha^{-1} + 20 kg N ha^{-1} + Manure 2.5 t ha^{-1} and manure 5 t ha^{-1}) thus giving a total of 36 treatments. They were laid out in a Partially Balanced Incomplete Block Design (PBIBD) with six incomplete blocks per replicate each containing six treatments, replicated 3 times making a total of 108 plots. Treatments were assigned to blocks randomly with plot size of 6 m × 4 m. The dry land sorghum (Gadam) and cowpea (M66) varieties were used as the test crops. Then at the end of the short rain 2011 season, smallholder farmers were invited for a field day to evaluate each plot by scoring in a scale of good, fair and poor according to their own observation on crop performance and this was compared with scientific data collected on crop productivity. They were all given equal opportunity to evaluate 108 plots in the field experiment. They were also asked the kind of water harvesting and soil fertility management they used in their farms.

Data Analysis

Social data was coded and analyzed with SPSS version 17. Data was analyzed by use of descriptive analysis where frequencies of scores for each treatment were computed. Dependency tests were also conducted to find out if there was a relationship between gender and the treatment score. The biophysical data on crop yield was analyzed using statistical Analysis of Variance (ANOVA) using SAS version 8. Differences between treatment effects were declared significant at $P \leq 0.05$.

Results

Farmer's Evaluation on Treatment Performance

In Kiritiri division, the farmers' criteria for distinguishing plots was on a scale of good, fair and poor that included crop yield and performance. The findings (Table 16.1) underscore the value of taking into consideration the visual and morphological crop characteristics used by farmers as a key criterion for scientific crop evaluation and development during Long rains 2011, 2012 and short rain 2011.

Table 16.1 Farmers' rating of water harvesting, cropping systems and ISFM technologies

Water harvesting	Cropping systems	Soil fertility management regimes	Mean score	Mean rank	Overall rating
Tied ridges	Sole crop	40 Kg P ha^{-1} + 20 Kg N ha^{-1} + Manure 2.5 t ha^{-1}	2.9	2,873.87	Good
Contour furrows	Sole crop	40 kg P ha^{-1} + 20 kg N ha^{-1} + Manure 2.5 t ha^{-1}	2.8	2,787.23	Good
Tied ridges	Intercrop	40 kg P ha^{-1} + 20 kg N ha^{-1} + Manure 2.5 t ha^{-1}	2.74	2,763.39	Good
Contour furrows	Intercrop	40 kg P ha^{-1} + 20 kg N ha^{-1} + Manure 2.5 t ha^{-1}	2.65	2,753.45	Good
Tied ridges	Sole crop	40 kg P ha^{-1} + 20 kg N ha^{-1}	2.64	2,677.23	Good
Contour furrows	Sole crop	Manure 5 t ha^{-1}	2.63	2,621.8	Good
Tied ridges	Sole crop	40 kg P ha^{-1} + 40 kg N ha^{-1} + Manure 5 t ha^{-1}	2.61	2,560.8	Good
Tied ridges	Sole crop	40 kg P ha^{-1} + 40 kg N ha^{-1}	2.53	2,490.58	Good
Contour furrows	Sole crop	40 kg P ha^{-1} + 40 kg N ha^{-1} + Manure 5 t ha^{-1}	2.51	2,443.79	Good
Tied ridges	Intercrop	40 kg P ha^{-1} + 40 kg N ha^{-1} + Manure 5 t ha^{-1}	2.49	2,462.85	Fair
Tied ridges	Intercrop	40 kg P ha^{-1} + 40 kg N ha^{-1}	2.48	2,403.52	Fair
Contour furrows	Sole crop	40 kg P ha^{-1} + 40 kg N ha^{-1}	2.47	2,384.7	Fair
Contour furrows	Sole crop	40 kg P ha^{-1} + 20 kg N ha^{-1}	2.45	2,397.72	Fair
Tied ridges	Sole crop	Manure 5 t ha^{-1}	2.44	2,310.8	Fair
Tied ridges	Intercrop	40 kg P ha^{-1} + 20 kg N ha^{-1}	2.43	2,368.56	Fair
Contour furrows	Intercrop	Manure 5 t ha^{-1}	2.43	2,338.03	Fair
Contour furrows	Intercrop	40 kg P ha^{-1} + 20 kg N ha^{-1}	2.38	2,280.98	Fair
Contour furrows	Intercrop	40 kg P ha^{-1} + 40 kg N ha^{-1} + Manure 5 t ha^{-1}	2.37	2,243.88	Fair
Contour furrows	Intercrop	40 kg P ha^{-1} + 40 kg N ha^{-1}	2.37	2,256.27	Fair
Tied ridges	Intercrop	Manure 5 t ha^{-1}	2.35	2,248.08	Fair
Farmers practice	Intercrop	40 kg P ha^{-1} + 20 kg N ha^{-1} + Manure 2.5 t ha^{-1}	2.32	2,170.3	Fair
Farmers practice	Sole crop	40 kg P ha^{-1} + 20 kg N ha^{-1}	2.31	2,145.78	Fair
Farmers practice	Sole crop	40 kg P ha^{-1} + 40 kg N ha^{-1}	2.31	2,138.53	Fair
Farmers practice	Sole crop	40 kg P ha^{-1} + 20 kg N ha^{-1} + Manure 2.5 t ha^{-1}	2.29	2,115.15	Fair

(continued)

Table 16.1 (continued)

Water harvesting	Cropping systems	Soil fertility management regimes	Mean score	Mean rank	Overall rating
Farmers practice	Intercrop	40 kg P ha^{-1} + 40 kg N ha^{-1}	2.28	2,102.65	Fair
Farmers practice	Sole crop	40 kg P ha^{-1} + 40 kg N ha^{-1} + Manure 5 t ha^{-1}	2.27	2,075.92	Fair
Farmers practice	Sole crop	Manure 5 t ha^{-1}	2.23	1,765.56	Fair
Farmers practice	Intercrop	40 kg P ha^{-1} + 40 kg N ha^{-1} + Manure 5 t ha^{-1}	2.02	1,621.56	Fair
Farmers practice	Intercrop	Manure 5 t ha^{-1}	2.0	1,615.85	Fair
Farmers practice	Intercrop	40 kg P ha^{-1} + 20 kg N ha^{-1}	1.74	1,524.25	Fair
Tied ridges	Sole crop	Control	1.43	1,101.24	Poor
Tied ridges	Intercrop	Control	1.43	1,095.52	Poor
Contour furrows	Sole crop	Control	1.42	1,085.25	Poor
Contour furrows	Intercrop	Control	1.38	954.2	Poor
Farmers practice	Sole crop	Control	1.19	658.86	Poor
Farmers practice	Intercrop	Control	1.03	568.27	Poor

(N = 366), *Test statistics* Kruskal-H test; Chi-square = 1,212.6; d.f = 35; $p = 0.000$

The results in Table 16.1 show that treatments under tied ridges with sorghum alone plus soil amendment of 40 kg P ha^{-1} + 20 kg N ha^{-1} + Manure 2.5 t ha^{-1} attracted the highest preference of farmers who rated it as 'good' with a mean score of 2.9 and was ranked number one out of 36 treatments. The experimental results (Table 16.2) also indicated that the same treatment had the highest amount of grain yield (3.7 t ha^{-1}). This was followed closely by contour furrows under the same cropping system and soil fertility amendment option (Tables 16.1 and 16.2) rated as 'good' with grain yield (3.5 t ha^{-1}). The results (Tables 16.1 and 16.2) further shows that all the treatments rated as 'Good' by the farmers were also the highest in grain yield ranging from 2.7 to 3.7 t ha^{-1}. The results show that all the technologies ranked 'good' included minimal combination of fertilizers and manure, or stand alone fertilizer application. However, the treatment which was rated by majority farmers as 'poor' was experiment control under farmers practice with sorghum and cowpea intercrop with a mean score of (1.03), mean rank (568.27) and yielding as low as (0.4 t ha^{-1}). Generally, all experiment controls were overall scored as 'poor' yielding as low as 0.4–0.7 t ha^{-1}.

Treatment Score by Gender

The results indicated that there was no significant difference ($P \geq 0.05$) regarding scoring by gender groups in all the 36 treatments of experiment which were ranked in the scale of good, fair and poor. However, there was a highly significant difference ($p < 0.001$) on rating of treatments by smallholder farmers in Mbeere District.

Field Experiment Results

The results underscore the scientific crop evaluation from the field experiment during 2012 Long rains. The results in Table 16.2 show performance of three types of water harvesting, two cropping system and six fertility levels but only that differed significantly from one another ($p = 0.0001$) in terms of sorghum grain yield. The three levels of water harvesting and the two cropping systems did not differ significantly in terms of grain yield among themselves ($p = 0.8413$) and ($p = 0.7168$) respectively. The total dry matter amount varied significantly among levels of cropping system and fertilizer application ($p = 0.0216$ and 0.0001) respectively. However the total dry matter amount did not vary significantly across water harvesting methods ($p = 0.5743$). The sorghum biomass were significantly different among cropping system ($p = 0.0020$) while water harvesting and fertility levels did not differ significantly ($p = 0.3930$ and 0.0698).

Table 16.2 The effects of water harvesting, cropping system and soil fertility regimes on sorghum yields in Kiritiri division

Water harvesting	Cropping system	Soil fertility management regimes	Stover + husks (t ha^{-1})	Grain yield (t ha^{-1})
Tied ridges	Sole crop	40 kg P ha^{-1} + 20 kg N ha^{-1}+Manure 2.5 t ha^{-1}	3.5	3.7
Contour furrows	Sole crop	40 kg P ha^{-1} + 20 kg N ha^{-1}+Manure 2.5 t ha^{-1}	3.5	3.5
Tied ridges	Intercrop	40 kg P ha^{-1} + 20 kg N ha^{-1}+Manure 2.5 t ha^{-1}	3.5	3.1
Contour furrows	Intercrop	40 kg P ha^{-1} + 20 kg N ha^{-1} + Manure 2.5 t ha^{-1}	3.4	3.1
Tied ridges	Sole crop	40 kg P ha^{-1} + 20 kg N ha^{-1}	3.4	3.1
Contour furrows	Sole crop	Manure 5 t ha^{-1}	3.5	2.9
Tied ridges	Sole crop	40 kg P ha^{-1} + 40 kg N ha^{-1} + Manure 5 t ha^{-1}	3.2	2.9
Tied ridges	Sole crop	40 kg P ha^{-1} + 40 kg N ha^{-1}	3.4	2.8
Contour furrows	Sole crop	40 kg P ha^{-1} + 40 kg N ha^{-1} + Manure 5 t ha^{-1}	3.2	2.7
Tied ridges	Intercrop	40 kg P ha^{-1} + 40 kg N ha^{-1} + Manure 5 t ha^{-1}	3.2	2.6
Contour furrows	Sole crop	40 kg P ha^{-1} + 40 kg N ha^{-1}	3.1	2.6
Contour furrows	Sole crop	40 kg P ha^{-1} + 20 kg N ha^{-1}	3.0	2.6
Tied ridges	Intercrop	40 kg P ha^{-1} + 20 kg N ha^{-1}	3.0	2.6
Contour furrows	Intercrop	40 kg P ha^{-1} + 40 kg N ha^{-1} + Manure 5 t ha^{-1}	3.0	2.5
Tied ridges	Intercrop	40 kg P ha^{-1} + 40 kg N ha^{-1}	3.4	2.5
Contour furrows	Intercrop	40 kg P ha^{-1} + 40 kg N ha^{-1}	3.3	2.5
Contour furrows	Intercrop	40 kg P ha^{-1} + 20 kg N ha^{-1}	3.3	2.4
Tied ridges	Sole crop	Manure 5 t ha^{-1}	3.2	2.4
Contour furrows	Intercrop	Manure 5 t ha^{-1}	3.2	2.4
Tied ridges	Intercrop	Manure 5 t ha^{-1}	3.2	2.3
Farmers practice	Intercrop	40 kg P ha^{-1} + 20 kg N ha^{-1} + Manure 2.5 t ha^{-1}	3.1	2.3
Farmers practice	Sole crop	40 kg P ha^{-1} + 20 kg N ha^{-1}	2.9	2.3
Farmers practice	Sole crop	40 kg P ha^{-1} + 40 kg N ha^{-1}	3.5	2.2
Farmers practice	Sole crop	40 kg P ha^{-1} + 20 kg N ha^{-1} + Manure 2.5 t ha^{-1}	3.6	2.2

(continued)

Table 16.2 (continued)

Water harvesting	Cropping system	Soil fertility management regimes	Stover + husks (t ha^{-1})	Grain yield (t ha^{-1})
Farmers practice	Intercrop	40 kg P ha^{-1} + 40 kg N ha^{-1}	3.2	2.2
Farmers practice	Intercrop	40 kg P ha^{-1} + 20 kg N ha^{-1}	3.3	2.2
Farmers practice	Sole crop	40 kg P ha^{-1} + 40 kg N ha^{-1} + Manure 5 t ha^{-1}	2.9	2.2
Farmers practice	Intercrop	40 kg P ha^{-1} + 40 kg N ha^{-1} + Manure 5 t ha^{-1}	4.0	2.1
Farmers practice	Intercrop	Manure 5 t ha^{-1}	3.9	2.1
Farmers practice	Sole crop	Manure 5 t ha^{-1}	3.7	2.0
Tied ridges	Sole crop	Control	1.2	0.7
Tied ridges	Intercrop	Control	0.9	0.7
Contour furrows	Sole crop	Control	1.3	0.7
Contour furrows	Intercrop	Control	1.8	0.6
Farmers practice	Sole crop	Control	1.5	0.5
Farmers practice	Intercrop	Control	0.8	0.4
Means			2.9	2.2
CV			24.8	22.4
LSD			1.51	0.85

Combination Effect

The results further indicated that sorghum without manure application did not differ significantly in yield with treatments that did not receive fertilizer application. However, plots that received fertilizer and no manure gave slightly higher sorghum yield as compared to plots that received manure and no fertilizer (Table 16.2). The highest sorghum yield (3.7 t ha^{-1}) was recorded from tied ridges under sole sorghum cropping system with application of 40 kg P ha^{-1} + 20 kg N ha^{-1} + Manure 2.5 t ha^{-1}, followed by 3.5 t ha^{-1} under contour furrow under the same soil amendment practice. In the third place were three treatments (3.1 t ha^{-1}) under tied ridges and contour furrow and the top seven treatments yield did not differ significantly from one another ($p < 0.05$). The lowest sorghum yield (<2.0 t ha^{-1}) was observed in treatments regarded as 'control' with neither fertilizer nor manure regardless of other intervention (water harvesting methods or cropping systems). The total dry matter and biomass were highest in tied ridges under sole cropping of soil fertility amendment of 40 kg P ha^{-1} + 20 kg N ha^{-1} + Manure 2.5 t ha^{-1} (7.7 t ha^{-1}) and (3.5 t ha^{-1}) respectively. All these top producers did not differ significantly from one another ($p < 0.05$) except from the experiment controls.

Discussions

Farmer's Evaluation on Treatment Performance

The consistently high preference (Table 16.1) by farmers on overall rating as 'good' and high grain yields (3.7 t ha^{-1}) on tied ridges and contour furrow under sorghum alone with a minimum combination of organic and inorganic inputs at half dose application of Nitrogen and manure. This was an indication that minimal nutrient replenishment was required in all the season in Mbeere district. Studies by Mugendi et al. (2010) and Gachimbi (2002) have also reported that farms in Mbeere require nutrient replenishment every season from manures, fertilizers and from of crop residue return in their farms. It has also been reported by Njeru et al. (2009, 2010, 2013) and Mairura et al. (2007) that soil fertility can be accessed through visual observation on crop performance and yield and therefore farmers were able to evaluate all the treatments to their level best. The results (Tables 16.1 and 16.2) further shown that water harvesting technologies that integrates soil fertility management technologies played a major role in moisture conservation and increased crop productivity and were also ranked highly by the farmers. This is in agreement with what Miriti et al. (2012) has further found that farmer perception on soil fertility is closely related to the soil's water holding capacity.

The results (Tables 16.1 and 16.2) shows that the third and the fourth treatments of tied ridges and contour furrow under sorghum and cowpea intercrop with the same soil fertility management options were dominated by their sole cropping systems. This could be a result of nutrient competition since cowpeas are heavy nutrient miners as they are associated with interspecific competition in mixed stands. The same results have been reported by Katsaruware and Manyanhaire (2009) that crop yield reduction can be experienced in intercrops where they are associated with interspecific competition in mixed stands and the absence of interspecific competition in the monocrops. The results further indicate that probably intercropping sorghum with cowpea depressed sorghum yields and this influenced farmer's decision on crop performance. This outcome for sorghum (Tables 16.1 and 16.2) could be in line with reports for maize from Kenya (Nadar 1984) and in Tanzania (Jensen et al. 2003) where maize grain yields reduction of 46–57 and 9 % occurred when maize was intercropped with cowpea due to the competition for moisture between the two crops. Alternatively due to slow mineralization of manure which needed a number of seasons to meet the level of nutrient competition (Lekasi et al. 2003). The results by Miriti (2011) have also shown that cowpea was also a nutrient competitor for maize production in semi-arid areas of eastern Kenya. Therefore, the results had a very clear relationship on their comparison on farmer's perception and crop yield. However, all those treatments regarded as 'controls' were poorly rated by the farmers and they had lower crop yields. The farmers practice under sorghum and cowpea intercrop were rated as 'poorly' with the lowest grain yield. This is in line with continuous cultivation of the same piece of land as this will lead to nutrient depletion and requires nutrient replenishment (Mugwe et al. 2009; Miriti et al. 2003). This has led to land degradation contributing to reduced crop production as a result of failure of rainfall distribution in semi-arid areas of Embu County. The farmers are being discouraged from adoption of these water conservation structures as a result of labour shortage and land tenure uncertainty (Demelash and Stahr 2010). Therefore, land productivity can be improved by employing of appropriate agricultural technologies which suit these semi-arid areas of Mbeere south District, Embu County.

Conclusions

The results reported in the study demonstrate that smallholder farmers' knowledge can provide a consistent treatment evaluation as compared to biophysical data. This demonstrated clear evidence from the study that there was a relationship of treatment rating by farmers with the scientific findings. However, there was no difference noted in terms of scoring of treatments by gender. Therefore, both genders could be used by agricultural extension services and researchers to evaluate other related scientific work in this study area. Mbeere south district is characterized by low and erratic rainfall and generally fragile ecosystems which are not suitable for sustainable rain-fed agriculture. The results have demonstrated the need to incorporate selected water harvesting and integrated soil fertility management technologies on sorghum and cowpea production in the season under low rainfall distribution in semi-arid areas. This will also suggests that only low-input technologies are currently suitable and need to be adopted through a known crop intensification technologies that could be enhanced in these areas. The results have also demonstrated a very clear message to smallholder farmer, extension services and other stakeholders that there is need for water harvesting technologies and nutrient replenishment on-farm every season to increase sorghum and cowpea productivity.

Acknowledgments This work was supported by International Development Research Centre (IDRC—Grant Number 105790-003) and Arid and Semi-arid Lands-Agricultural Productivity Research Project (ASAL-APRP). We would like to acknowledge the Director KARI, Centre Director KARI-Muguga and Kenyatta University supervisors for their facilitation. Authors would also like to acknowledge the contributions by staff from KARI-Muguga south, KARI-headquarters, KARI-Embu and KARI-Njoro for their assistance. We are grateful to all the area agricultural extension officers and farmers who participate in this work.

References

African Fertilizer Summit (2006) In: African fertilizer summit proceedings, IFDC, Muscle Shoals, p 182

Bationo A, Kimetu J, Ikeera S, Kimani S, Mugendi D, Odendo M, Silver M, Swift MJ, Sanging N (2004) The African network for soil biology and fertility: new challenges and opportunities. In: Bationo A (ed) Managing nutrient cycles to sustain soil fertility in Sub-Saharan Africa. Academy Science Publishers and Tropical Soil Biology and Fertility Institute of CIAT, Nairobi, Kenya, pp 1–23

Bielders CL, Michels K, Bationo A (2002) On-farm evaluation of ridging and residue management options in a Sahelian millet cowpea intercrop. Soil quality changes. Soil Use Manag 18:216–222

CAADP (2008) Agricultural policy-making in sub Saharan Africa. CAADP Progress in Kenya

Crews ET, Peoples BM (2004) Legume versus fertilizer sources of nitrogen: ecological tradeoffs and human needs. Agric Ecosyst Environ 102:279–297

Demelash M, Stahr K (2010) Assessment of integrated soil and water conservation measures on key soil properties in South Gonder, North-Western Highlands of Ethiopia. Soil Sci Environ Manag 1:164–176

Fongod AGN, Mih AM, Nkwatoh TN (2012) Morphological and agronomic characterization of different accessions of sweet potatoes (*Ipomoea batatas*) in Cameroon. Int Res J Agric Sci Soil Sci 2:234–245

Gachimbi LN (2002) Technical report of soil survey and sampling results: Embu-Mbeere Districts, Kenya. LUCID working paper series number: 9. KARI-kabete, Nairobi, Kenya

GoK (2007) Government of Kenya. Ministry of planning and national development. The Kenya Vision 2030 A competitative and Prosperous Kenya. Nairobi, Kenya

Jaetzold R, Schmidt H, Hornet ZB, Shisanya CA (2007) Farm management handbook of kenya. In: Natural conditions and farm information. 2nd edn. vol 11/C. Eastern Province. Ministry of Agriculture/GTZ, Nairobi, Kenya

Jensen JR, Bernhard RH, Hansen S, McDonagh J, MØberg JP, Nielsen NE, Nordbo E (2003) Productivity in maize based cropping systems under various soil-water-nutrient management strategies in semi-arid Alfisol environment in East Africa. Agric Water Manag 59:217–237

KARI (2009) Kenya Agricultural Research Institute. Strategic plan Implementation framework. (2009–2014). In: A KARI report. Nairobi, Kenya

Katsaruware RD, Manyanhaire IO (2009) Maize-cowpea intercropping and weed agronomy and soil science. Suppression in leaf stripped and detasselled maize in Zimbabwe. Electron J Environ Agric Food Chem 8:1218–1226 (Madison, WI, USA)

Kimani SK, Esilaba AO, Odera MM, Kimenye L, Vanlauwe BJ, Batiano A (2007) The effect of organic sources of nutrients on maize yield in three districts of central Kenya. In: Batiano A (ed) Advances in integrated soil fertility management in Sub-Sahara Africa, challenges and opportunities. Springer, Berlin, pp 283–288

Lekasi JK, Tanner JC, Kimani SK, Harris PJC (2003) Cattle manure quality in Maragua district, Central Kenya: effect of management practices and development of simple method of assessment. Agric Ecosyst Environ 94:289–298

Mairura FS, Mugendi DN, Mwanje JI, Ramisch JJ, Mbugua PK, Chianu JN (2007) Integrating scientific and farmers' evaluation of soil quality indicators in Central Kenya. Geoderma 139:134–143

Mbogoh S (2000) Makueni District profile: crop production and marketing. Dryland research working paper 7, Dryland Research, Crewkerne, UK, pp 1988–1999

Miriti JM (2011) Effect of conservation tillage on water and nitrogen use efficiency in maize-cowpea cropping systems in semi-arid eastern Kenya. Ph.D thesis, University of Nairobi, Kenya

Miriti JM, Kironchi G, Esilaba AO, Heng LK, Gachene CKK, Mwangi DM (2012) Yield and water use efficiencies of maize and cowpea as affected by tillage and cropping systems in semi-arid Eastern Kenya. Agric Water Manag 115:48–155

Miriti JM, Thomas DB, Gachene CCK (2003) Soil chemical properties, sunflower growth and yields as affected by tillage, compost and mulch application. E Afr Agric For J 68:159–163

Mugendi D, Mucheru-Muna M, Mugwe J, Bationo A (2003) Enhancement of food productivity using leguminous shrubs in the eastern highlands of Kenya. In: African crop science conference proceedings, vol 6, pp 581–586

Mugendi E, Gitonga N, Cheruiyot R, Maingi J (2010) Biological nitrogen fixation by promiscous sobean (Glycine max L. Merril) in the central highlands of Kenya. Response to soil amendends. World J Agric Sci 6:381–387

Mugwe J, Mugendi D, Mucheru-Muna M, Merckx R, Chianu J, Vanlauwe B (2009) Determinants of the decision to adopt integrated soil fertility management practices by smallholder farmers in the central highlands of Kenya. Expl Agric 45:61–75

Nadar HM (1984) Intercropping and intercrop component interaction under varying rainfall conditions in eastern Kenya. II: maize/cowpea intercrop. E Afri Agri For J 44:176–181

Njeru PNM (2009) The effect of push-pull management and different fertilisation strategies on maize crop yield in central Kenya. Msc. thesis, SLU—Uppsala University, Sweden

Njeru PNM, Hjalten J, Kimani SK (2010) A book on the effect of push-pull management and different fertilization strategies on maize crop yield in central Kenya. Lambert Acadamic publishers, German. SSBN:978-3-8433-6376-1

Njeru PNM, Maina I, Miruka M, Amboga JS, Gitari J, Rono B, Mahasi M, Murithi F (2011a) Soil fertility characterization of smallholder farms under group and individual management in central Kenya. In: African crop science conference proceedings, vol 10, Maputo, Mozambique, 10–13 Oct 2011, pp 1–4

Njeru PNM, Maina I, Mugwe J, Mugendi D, Miruka M, Lekasi JK, Kimani SK, Amboga JS, Gitari J, Mahasi M, Mwangi H, Muriithi F (2011b) Capacity development in soil science knowledge for increased food production on smallholder farms in Central Kenya. In: 16th soil science society of East Africa. Jinja, 22–26 Nov 2011, p 12

Njeru PNM, Mugwe J, Maina I, Mucheru-Muna M, Mugendi D, Lekasi JK, Kimani SK, Miriti J, Esilaba AO, Muriithi F (2013) Integrating scientific and farmers' perception towards evaluation of rain-fed agricultural technologies for sorghum and cowpea productivity in Central Kenya. J Soil Sci Environ Manag 4(7):123–131

Rangwa PK, Kamau NR, Mbatia ED (1998) Irrigation and dwaterage Branch position paper. Paper presented during the workshop on promotion of sustainable smallholder irrigation development in Kenya, Embu, Kenya, 16–20 Nov 1998

Sanginga N, Woomer PL (eds) (2009) Integrated soil fertility management in Africa: principles, practices and developmental process. Nairobi. Tropical soil biology and fertility institute of the international centre for tropical agriculture, p 263

Chapter 17
On-Station Evaluation of Maize Genotypes for Nutrient and Water Use Efficiency in the Semi Arid Lands of Coastal Kenya

F.N. Pole, H.M. Saha, N. Mangale, A.M. Mzingirwa and P. Munyambu

Abstract Evaluation of six maize varieties under four different water harvesting and tillage technologies was undertaken with the aim of determining their effect on the performance of maize genotypes and their effectiveness in improving nutrient and water use efficiency. The work was carried out at Mariakani site (one of the KARI centres) representing the arid and semi-arid lands of coastal Kenya in the long rains and short rains seasons of 2005 and 2006 respectively. The results indicate that rainwater harvesting is not critical when the season is wetter than normal in the arid and semi-arid environments. This was demonstrated by the high yields that were recorded from the maize varieties (Pwani Hybrid 4-PH4, Coast Composite Maize-CCM and the local check-Mdzihana) which usually require relatively high rainfall amounts in order for them to produce better yields. Despite the excellent performance of PH4, CCM and Mdzihana, these maize varieties cannot be recommended for the semi-arid areas since the high yields were realized under above normal rainfall. There is need for further research to identify the maize varieties that would be appropriate for the areas that normally receive low rainfall.

Keywords Arid and semi-arid areas · Crop failure · Low rainfall · Maize varieties · Water harvesting method

F.N. Pole (✉) · A.M. Mzingirwa · P. Munyambu
Kenya Agricultural Research Institute—Matuga, P.O. Box 4, Matuga 80406, Kenya
e-mail: finyange@gmail.com

H.M. Saha
Pwani University, P.O. Box 195, Kilifi 80108, Kenya

N. Mangale
Kenya Agricultural Research Institute, NARL, P.O. Box 14733, Westlands, Nairobi 00800, Kenya

W. Leal Filho et al. (eds.), *Adapting African Agriculture to Climate Change*, Climate Change Management, DOI 10.1007/978-3-319-13000-2_17

Introduction

Soil fertility management in semi-arid areas is mainly constrained by inadequate moisture and lack of resources for the purchase of inputs, particularly very expensive inorganic fertilizers that are usually imported. This problem is usually compounded by the nutrient removals which in turn exacerbate the problem of soil fertility decline. The nutrient removals arise from continuous cropping without adequate replenishment, soil erosion, and failure by farmers to judiciously manage soil nutrient reserves. Continuous cultivation of land exposes soil organic matter (SOM) to oxidative processes, since in most cases there is reduced biomass input (Shepherd et al. 1996). Decomposition of organic matter has an impact on the capacity of soil to retain moisture and nutrients. Soil organic matter plays an important role in soil fertility replenishment but the linkages between degradation and soil carbon sequestration and nutrient retention are complex.

Some forms of tillage practices, particularly in arid and semi-arid areas encourage oxidation of organic matter throughout the profile resulting in the release of carbon dioxide to the atmosphere rather than its build up in the soil (Nabhan and Buchmann 1997). This leads to reduced biomass production from crops and pastures, as well as lower carbon inputs to the soil in subsequent periods because low amounts of root matter, leaf litter and crop residues are returned to the soil. Many of the soils in the semi-arid zones of Eastern and Southern Africa are deficient in some essential mineral nutrients especially Nitrogen (N) and Phosphorus (P) (Okalebo et al. 2002). Low soil fertility, especially N and P deficiency, is a major biophysical constraint for successful agriculture in the semi-arid areas of East and Southern Africa (Yates 1992). A better understanding of nutrient imbalances induced by pH and applied nutrients is therefore important in managing soil fertility.

In Kenya, the semi-arid areas receive very low bimodal rainfall, with recordings as low as 342 mm per annum (Nadar and Faught 1990). The probability that the rainfall is less than two thirds of the potential evaporation in the rainy season varies between 60 and 80 %, equaling to two thirds of the potential evaporation (approximately 347 mm in a 4 months growing season). This is the minimum required by annual crops like maize and beans with a growing period of 3 months. Rainfall amounts that are less than half of the potential evaporation (approximately 275 mm within 4 months) during the growing season will lead to crop failure (Nadar and Faught 1990). The probability of crop failure in semi-arid regions may therefore be quite high in years with below-average rainfall.

Farmers in the East and Southern African region use various water capturing strategies aimed at reducing run-off and/or enhance infiltration. These include terracing on steep slopes particularly but not exclusively in Kenya, contour bands including dead level contours, storm drains and other water harvesting techniques such as tied ridges and basins (Zai pits) constructed in a field of growing crops to encourage water retention and infiltration. Tiffen et al. (1994) have reported the excellent progress made in Kenya (Machakos) in reducing soil erosion through contour banks and terracing. Terracing of the croplands using hand dug contour trenches and banks was long adopted by farmers in Kenya and is popularly known as ‘fanya juu’ terraces (Simpson et al. 1996). Tiffen et al. (1994) further affirms that any deleterious effects of changes in the nature of the base (due to cultivation and cropping) have been more than offset by improvements due to terracing and other conservation measures and that farmers have learned to manage the resources better. The conservation of soil and water has led to an increased supply of soil moisture and hence increased the potential for the mining of the remaining soil nutrients by crops (Simpson et al. 1996). With inherently low soil fertility, most of the soils in the semi-arid areas of East and Southern Africa may produce short term gains in crop yields when appropriate water harvesting techniques are used but will accentuate the problem of low soil fertility in the long term.

Optimization of soil water use for increased crop production in dry land agriculture is dependent on agronomic practices that result in an increase in the amount of water stored in the soil and efficient utilization of that water. This therefore calls for the need for crop genotypes that can utilize the limited water resources to sustain their growth for the benefit of mankind. The crop genotypes should also have the ability to efficiently utilize the lower amounts of nutrient available in those soils since farmers in the arid and semi-arid areas are resource poor and therefore, cannot afford expensive inorganic fertilizers. The two major pathways of water loss are evaporation from the soil surface and runoff. Losses due to evaporation can be as high as 66 % of the season’s rainfall (Smets et al. 2011). This brings in the idea of carrying out some tillage practices that can conserve water and reduce the rate of evaporation as well as runoff.

Coastal Kenya is located between latitudes 10° and 4° south and longitudes 38° and 41° east. It covers an area of approximately 84,000 km^2 and is subdivided into seven administrative areas namely Kilifi, Kwale, Mombasa, Malindi, Tana River, Lamu, and Taita Taveta. The largest part of the region lies in the coastal lowlands (CL) which is subdivided into five agro-ecological zones (AEZ) namely CL2, CL3, CL4, CL5 and CL6. The high potential areas are comprised of CL2 and CL4, while CL5 and CL6 form the arid and semi-arid lands (ASAL). The region receives a bimodal rainfall with annual averages ranging from 1,400 mm in CL2 to less than 400 mm in CL6. Rainfall is distributed over two distinct seasons; the long rains (April–July) and the short rains (October–December). The most common food crops grown in the region are maize, cassava and cowpeas. The average on-farm yield of maize is about one tone of grain per hectare. The region is therefore a food deficit area, producing only 20 % of its food requirements. Agricultural production in the coastal region is constrained by low inherent fertility of the soils and their poor water retention. The problems associated with low soil fertility and low water holding capacity have been aggravated by practices that cause net losses of nutrients and soil water. Increasing population pressure has resulted in reduced farm sizes and more intensive farming in the high potential areas (CL3 and CL4) without substantial replenishment of plant nutrients. Farms in these areas cannot be left fallow as was the tradition in the past. This has led to soil nutrient depletion and low crop yields.

In order to address the problem of food deficit in the region, there is a need to open up more land for agriculture. The arid and semi-arid areas of Coastal lowland Kenya could therefore offer an alternative site to supplement the high potential areas since they form 75 % of the coastal region. However, rainfall in the arid and semi-arid areas is usually low and erratic and unevenly distributed. In most cases the total amount of seasonal rainfall can occur in just a few days leading to both soil and water losses. Crop yields are extremely low due to not only inadequate soil moisture but also low soil fertility. Despite the situation, farmers persistently grow maize every season (Mangale et al. 2003). This calls for integrated technological interventions so as to make proper use of the low rainfall amounts in these areas for crop growth. The main objective of this study was to develop technologies for enhancing food production in the semi-arid areas of coastal lowland Kenya.

Materials and Methods

Study Area

The study was conducted on station at Kenya Agricultural Research Institute (KARI) Mariakani, Kilifi County of coastal Kenya for two cropping seasons in 2005 and one season in 2006. The site is 200 m above sea level. Soils at the study site are sandy loams. They are deep, well drained and have a fairly high water holding capacity. The rainfall pattern at the site is bimodal with peaks in May and November during the LR and SR seasons, respectively. The average annual rainfall is 500 mm. The mean monthly maximum and minimum temperatures are 35 °C and 28 °C, respectively.

Experimental Procedures

Land was prepared manually to establish the various water conservation structures after bush clearing and removal of stamps. A plot measuring 65 × 50 m was subdivided into 3 blocks of 19 × 48 m separated by a 2 m path. The blocks were further sub-divided into 4 sub-blocks measuring 10.8 × 19 m separated by a path of 1 m between them. Each sub-block was again sub-divided into 6 plots (3 from either side) measuring 3.6 × 9 m. The sub-blocks were randomly assigned the water-conservation methods and all the plots within the sub-block were prepared according to the particular water conservation method. The water harvesting technologies that were tested include; Tumbukiza or Zai pits (60 cm long × 60 cm wide × 15 cm deep), deep tillage (using the long handle big hoe or oxen plough) and tied ridges. The farmer practice of shallow tillage was used as a control. Six maize genotypes (PH1, DHO2, CLS3, PH4, CCM, and a Local check) were planted and tested for water use efficiency. All the plots were applied with farm yard manure (FYM) at the rate of 5 tons per hectare. A randomized complete block experimental (RCBD) was used, with a split-plot arrangement of treatments, with water harvesting technologies forming the main plots while the maize genotypes formed the sub-plots.

Data Analysis

The recorded data was subjected to the analysis of variance (ANOVA) using the statistical package SAS. Significance was tested at the 5 % level. Where there were significant differences, the means were separated using least significant differences (LSD).

Results and Discussion

The first season's experiment (2005 LR) was characterized by a moderate amount of rainfall that saw the crop perform quite well. However, a few weeks before harvesting, it was severely damaged by rogue elephants that had strayed from the nearby game reserve. This made it difficult to obtain the yield data. The second experiment, which was conducted during the 2005 SR season, received very low amounts of rainfall which resulted in the crop drying up after flowering stage. Data was therefore collected and analyzed for plant heights and maize stover yield.

The results showed that water harvesting method had significant ($p < 0.05$) effect on maize plant height and stover yield (Table 17.1). Zai pits and deep tillage increased the maize plant height by 19–20 % as compared to shallow tillage. Maize under deep tillage practice was 1.2 times taller than that under shallow tillage. Tied ridges on the other hand were not significantly superior to shallow tillage in their effect on maize plant height and stover yield.

Table 17.1 Effect of water harvesting technology on maize stover yield and plant height—2005 SR season

Water harvesting technology	Stover yield (t ha^{-1})	Plant height (cm)
Deep tillage	0.95^{a}	39.5^{a}
Tied ridges	0.84^{ab}	36.4^{ab}
Zai pits	0.84^{ab}	39.4^{a}
Shallow tillage (using traditional hoe)	0.71^{b}	33.0^{b}
$LSD_{0.05}$	0.178	3.71
CV (%)	31.8	21.4

Values within the column followed by the same superscript are not significantly different at 5% level

Results of the 2006 long rains indicate that water harvesting methods had no significant effect on grain yield (Table 17.2). This was probably due to the excessive amount of rainfall that was received at the experimental site during the months of May and June, to the extent that water was not a limiting factor. The excessive rains resulted in water being collected in the zai pits thereby leading to water logging for two to three days after rainfall events. Some of the pits were also filled with silt. The water logging conditions led to a reduction in the growth rate of maize in the zai pits. This explains the low stover yields in the zai pits than that realized from maize grown under tied ridges.

Table 17.3 shows that maize variety had significant ($P < 0.05$) effect on maize plant height and stover yield during the 2005 SR season. The dry land hybrid maize (DHO2) was 11, 39, 53 and 81 % taller than CCM, Mdzihana, PH1, PH4 and CLS3 respectively. The DHO2 maize seemed to grow faster than the rest of the maize genotypes but the high rate of growth was not reflected in the genotype's stover yield. The coastal hybrids (PH1 and PH4) were 20–27 % shorter than CCM but were at the same time not significantly different from the local maize (Mdzihana) in their heights. Mdzihana, PH1, CCM and DHO2 did not differ in their stover yields. Dryland hybrid maize (DH02) was 1.5 times as tall as the coastal hybrids (PH1 and PH4).

Table 17.2 Effect of water harvesting technology on maize stover yield and plant height during 2006 LR season

Water harvesting technology	Grain yield (t ha^{-1})	Stover yield (t ha^{-1})
Deep tillage	3.80	5.5^{b}
Tied ridges	3.95	6.5^{a}
Zai pits	3.67	5.6^{b}
Shallow tillage (using traditional hoe)	3.83	5.8^{ab}
$LSD_{0.05}$	NS	0.75
CV (%)	22.7	19.0

Values within the column followed by the same superscript are not significantly different at 5% level

Table 17.3 Effect of maize genotype on stover yield and plant height—2005 SR season

Maize genotype	Stover yield (t ha^{-1})	Plant height (cm)
Pwani hybrid 4 (PH4)	0.97^{a}	32.2^{c}
Coast composite maize (CCM)	0.93^{ab}	44.4^{b}
Dry land hybrid maize (DH02)	0.84^{abc}	49.3^{a}
Mdzihana (local maize)	0.83^{abc}	35.4^{c}
Pwani hybrid 1 (PH1)	0.75^{bc}	32.2^{c}
Coastal lowland synthetic 3 (CLS3)	0.70^{c}	27.2^{d}
$LSD_{0.05}$	0.218	4.55
CV (%)	31.8	21.4

Values within the column followed by the same superscript are not significantly different at 5% level

Table 17.4 Effect of maize genotype on stover yield and plant height—2006 LR season

Maize genotype	Stover yield (t ha^{-1})	Plant height (cm)
Pwani hybrid 4 (PH4)	4.5^{a}	7.8^{a}
Coast composite maize (CCM)	4.0^{ab}	6.4^{b}
Dry land hybrid maize (DH02)	3.0^{c}	3.0^{d}
Mdzihana (Local maize)	3.9^{ab}	8.4^{a}
Pwani hybrid 1 (PH1)	3.8^{b}	4.8^{c}
Coastal lowland synthetic 3 (CLS3)	3.7^{bc}	4.7^{c}
$LSD_{0.05}$	0.71	0.92
CV (%)	22.7	19.0

Values within the column followed by the same superscript are not significantly different at 5% level

Results of the 2006 LR season showed that maize genotype had a significant effect on maize grain and stover yield (Table 17.4). Pwani hybrid 4 (PH4), Coast Composite (CCM) and the local maize variety (Mdzihana) had higher grain and stover yields than DHO2, PH1 and CLS3. However, these maize varieties cannot be recommended for the semi-arid areas despite their excellent performance since their high yields were realized due to the above normal rainfall that was received at the site during the time of the experiment. There is need therefore to repeat the experiment at the same site with low moisture levels so as to determine the appropriate maize varieties for the semi-arid areas.

Recommendation and Way Forward

The results from the experiment indicate that rainwater harvesting for crop production in the semi-arid areas is not critical when the season is wetter than normal. Despite the excellent performance of PH4, CCM and Mdzihana, these maize varieties cannot be recommended for the semi-arid areas since the high yields were only realized after the region received above normal rainfall during that particular season. There is need for further experimentation so as to identify the varieties that are suitable for the area that is characterized by low amounts of rainfall.

References

Mangale N, Danda K, Muinga RW, Saha HM, Muli MB, Muniu FK, Wachira S (2003) Characterization of farmers and farming systems to establish entry points for soil and water management interventions to enhance food security in coastal Kenya. Paper presented at the 1st adaptive research conference, held at KARI HQ, June 2003

Nabhan GP, Buchmann SL (1997) Services provided by pollinators. In: Daily GC (ed) Natures Services. Societal Dependence of Natural Ecosystem. Island Press, Washington, DC, pp 133–150

Nadar HM, Faught WA (1990) Effect of legumes on the yield of associated and subsequent maize in intercropping and rotation systems without nitrogen fertilizer. East Afr Agric For J. 44 (Special Issue):127–136

Okalebo JR, Gathua KW, Woomer PL (2002) Laboratory methods of soil and plant analysis: a working manual, 2nd edn. TSBF-CIAT, SACRED Africa, SSSEA, KARI, and Moi University, Nairobi, Kenya. pp 128

Shephered KD, Ohlsson E, Okalebo JR, Ndufa KJ (1996) Potential impact of agroforestry on soil nutrient balances at farm scale in East African highlands. Fertilizer Res 44:87–89

Simpson JR, Okalebo JR, Lubulwa G (1996) The problem of maintaining soil fertility in Eastern Kenya: a review of relevent research. ACIAR (Australia), monograph no. 41, p 60

Smets T, López-Vicente M, Poesen J (2011) Impact of subsurface rock fragments on runoff and interrill soil loss from cultivated soils. Earth Surf Process Land J 36:1929–1937

Tiffen M, Mortimore M, Gichuki F (1994) More people less erosion. Environmental recovery in Kenya. ACTS. Nairobi, Kenya and ODI, London

Yates SR (1992) An analytical solution for one-dimensional transport in porous media with an exponential dispersion function. Water Resour Res 28:2149–2154

Chapter 18
Tomato (*Lycopersicon Esculentum* Mill.) Yield Performance under Elevated Dry Season Temperatures as an Adaptation to Climate Change in Tabora, Tanzania

Fabian M. Bagarama

Abstract Tomato growing is an adaptation strategy to rainfall variability and droughts that frequently result into loss of the maize crop during the rainy season. This study assesses yield performances of tomato genotypes under elevated dry season air temperatures in semi-arid environments as a climate change adaptive practice. Tomato is mostly grown in June through August months. The dry season is characterized by maximum daily temperatures (34.2 °C), low night temperatures (14.8 °C) and monthly evaporation of 168.2–226.6 mm between July and October. The atmospheric humidity is between 46 and 52 % in the same period. Highly significant yield at ($P = 0.01$) differences were found between the tested genotypes; Oxyl, Tanya Mkulima and Tengeru. Tomato planted in the month of August gave very low yields compared to the June planted crop. Low tomato yields were recorded on smallholders' farms. Infestation by red spider mites (*Tetranychus evansi)* increased with increasing temperatures and reduced irrigation. Application of NPK with secondary nutrients Ca, Mg, S, and Zn improved tomato yield under elevated air temperatures.

Keywords Agrometeorology · Dry season farming · Genotypes · Tanzania · Tomato

Introduction

Dry season vegetable growing is an important economic activity to farmers in some sub-Saharan African countries such as in Ghana (Nakuja et al. 2012), Nigeria (Ojo et al. 2011), Uganda (Ssekabembe et al. 2003) and Tanzania. Dry season vegetable growing is also considered by farmers as a climate change adaptation

F.M. Bagarama (✉)
Ministry of Agriculture, Food Security and Cooperatives, Department of Research and Development, Tumbi Agricultural Research Institute, P. O. Box 306, Tabora, Tanzania
e-mail: bagaramaf@gmail.com

W. Leal Filho et al. (eds.), *Adapting African Agriculture to Climate Change*,
Climate Change Management, DOI 10.1007/978-3-319-13000-2_18

strategy to compensate for crop yield losses from high rainfall variability during the main season (Mongi et al. 2010). Constraints to dry season vegetable growing among others include water shortage (Ikilulu 2006; Tsoho and Salau 2012) and high incidences of pests and diseases (Tsoho and Salau 2012; Bagarama 2013). Dry season farming is even more constrained by access to water in areas where farmers mainly depend on underground water and small hand-dug reservoirs (IKilulu 2006). Heat stress is another major abiotic factor that limits tomato production during summer season. High air temperatures negatively affect plant growth and survival and hence crop yield (Boyer 1982). According to a recent study, each degree centigrade increase in average growing season temperature may reduce crop yield by up to 17 % (Lobell and Asner 2003). Lack of tolerance to high temperature in most tomato genotypes presents a major limitation for growing tomato where temperature during part of the growing season, even for short durations, reach 38 °C or higher (Abdul-Baki 1991). Maximum temperature in semi-arid Tabora occasionally reaches 35 °C in the dry season. For tomato, the 8–13 day period prior to anthesis is the most critical development phase (Higashide 2009). In tomato, elevated temperature impacts are complex, and it is difficult to determine one critical temperature effect during the reproductive development phase. In experiments conducted to determine critical temperature effects, sensitive cultivars are impacted when mean daily temperature exceed 25 °C, whereas more heat tolerant cultivars are not impacted until maximum temperature exceeded 32 °C. Varieties vary in their sensitivity to temperature and this will influence pollination and fruit set. Fruit setting is reduced when temperature rise above 27 °C. Optimum temperature for fruit set is 18–24 °C (Lovatt et al. 1998). Even moderate increases in daily temperatures (from 28/22 to 32/26 °C day/night) have been shown to result in a significant decrease in the number of fruit set (Peet et al. 1997; Sato et al. 2006). Degeneration of the embryo sac due lack of fertilization is one of the main reason for flower abscission (Ormrod et al. 1967). Pollen is more affected by heat stress than ovule (Monterroso and Wien 1990). Current studies show that there is close relationship between pollen sustainability and tolerance to high temperature stress among bean selections. Tomato plant growth demands adequate moisture 400–600 mm and soil nutrients especially the three macronutrients nitrogen, phosphorus and potassium (Landon 1984). It is forecasted that tropical countries will be affected most by climate change mostly reflected in terms of crop yield decline (Fischer et al. 2002). Water shortage and heat stress are some of the anticipated and clearly felt causes in Tanzania. The spatial and temporal impact of environmental variability is continually growing. This requires establishing resilience based research to increase immediately farmers' capacity towards responding to environmental stress.

Studies on climate change in western Tanzania have focused much on rainfall variability and farmers' vulnerability to rainfall variability (Mongi et al. 2010). Farmers in Tabora, region in western Tanzania grow tomato and green maize as an adaptation to climate change challenges (Mongi et al. 2010; Bagarama 2013). Currently there is no information on growth and performance and tolerance to high temperatures of tomato in western Tanzania. This information is vital to help farmers adopt tomato genotypes that are tolerant to higher air temperatures during the dry season.

Materials and Methods

Study Area

Field experiments were conducted at the Ministry of Agriculture's Tumbi Training Institute farm (S05°04′08.0″ E032°41′15.3″ Elevation 1210 m.a.s.l). The study area is located in the Miombo woodland ecosystem of Western Tanzania. The soil on the experimental site was the loam-clay soil. The site had been cleared of natural vegetation for dam construction but it was strongly under depositions from the upper slopes.

Climatic Conditions During Tomato Growing Period in Tabora, Region

Tomato is normally grown during the period July–October when atmospheric temperatures and evaporation are raising. Temperatures range from a minimum of 13.6 °C in June–July to a mean maximum of 32.5 °C in September–October. The average annual temperatures are 23 °C. This period also coincides with appreciable decline in water in the adjacent dam. Atmospheric temperatures, humidity and evaporation data from July–October were obtained from a nearby Meteorological station located about 600 m from the experimental site.

Field Experiments

Field trials were carried out with four tomato genotypes—Tanya Mkulima, Tengeru, Oxyl and Tanya Kiboko. The first study involved the evaluation of the effect of organic manures on yield of variety Tanya Mkulima. The experiment was a randomized complete block design in three replications. Tomato seedlings were planted on raised seed beds at the spacing of 60 cm by 45 cm on alluvial soil. The seedlings were fertilized using 250 gm per hill of well decomposed manure. Top dressing with calcium nitrate at the rate of 8 gm per plant was done two weeks after transplanting. Tomato seedlings were watered using underground water from nearby water wells. Tomato fungal diseases were controlled by spraying LINKMIL 72 WP a protective and curative broad spectrum fungicide. Spraying of tomato plants was done once per week. Three tomato varieties, Tengeru, Oxyl and Tanya Mkulima were evaluated in the second trial using complete randomized block design. The experiments were conducted over two seasons 2011–2012. Tomato seedlings were transplanted between June through August. On-farm assessment of performance of the selected tomato varieties under farmers' conditions was also carried out during the same period. On-farm studies are aimed at identification of tomato genotypes and the associated constraints to production. The associated late blight disease and red spider mites (*Tetranychus evansi* Baker and Prichard) infections was evaluated by counting the number of infected plants out of 40 sampled plants. The data was analyzed using the ANOVA statistical procedures.

Results of the Field Experiment

The results from the field experiment are shown in Tables 18.1, 18.2 and 18.3. The response of tomato to balanced application of NPK fertilizers with micronutrients is shown in Table 18.1. The influence of the time of tomato transplanting date on yield for the three tomato genotypes is in Table 18.2 and the effect of late (August) planting on the yield of tomato variety Tanya Kibo is in Table 18.3.

Table 18.1 The effect of NPK fertilizer application on the yield of June planted tomato (var Tanya Mkulima) in Tabora, Tanzania

Fertilizer type	Number of tomato fruits/plant
Control (cow manure + calium nitrate)	93
N10P18K24 (S, Zn)	128
N20P10K10 (S, Zn)	170
LSD0.05	32.8
CV%	14.5

Table 18.2 Tomato genotypes yield performance during dry season in Tabora, Tanzania

Tomato genotype	Early August planted	Mid August planted
Oxyl	31.3	16.6
Tanya Mkulima	31.9	18.9
Tengeru	12.7	6.8
LSD 0.05	10.75	5.16
CV%	19.8	39.9

Table 18.3 The yield of tomato genotype var Tanya Kibo planted in August, in Tabora, Tanzania

Date of planting	Number of fruits/plant
05/08/2012	22.1
21/08/2012	27.2
31/08/2012	26.5
LSD0.05	5.85
CV%	30.3

Table 18.4 On-farm tomato genotypes yield performance for Early August planted tomato genotypes in Tabora, Tanzania

Tomato genotype	Mean fruit/plant	Late blight disease infection (%)	Red spider mite infestation (%)
Tanya Mkulima	6.75	12.5	37.5
Grifforn	5.85	72.08	8.33
Anna	11.47	16.6	10.0

On-Farm Tomato Yield Performance Results

Results of on-farm performance of tomato varieties during dry season under high air temperatures are shown in Tables 18.4 and 18.5. Three tomato genotypes were found on farm during the period of the study. These were Anna, Grifford and Tanya Mkulima (Table 18.4).

Discussions

Results of this study show that there was a climate induced stress on tomato plants as the maximum temperatures between July–October were between 30.5 and 34.2 °C. The atmospheric humidity was low between 46 % in July and 52 % in October. Total monthly evaporation ranged from 168.2 mm in July to 226.6 mm in October. According to Lovatt et al. (1998) tomato fruit setting is reduced when temperatures rise above 27 °C and documented that the optimum temperature for fruit set is 18–24 °C. The increase in mean daily temperatures (from 28/22 to 32/26 °C day/night) have been shown to result in a significant decrease in the number of fruit set (Peet et al. 1997; Sato et al. 2006). In this study the maximum/minimum temperature range between July and October was 33.1/18.2 °C recorded in September and 30.5/14.8 °C recorded in July. The climate is thus characterized by high day temperatures and low night temperatures. There was no significant difference on the number of tomato fruits per plant for the genotype Tanya Kibo when planted late on August 05, 21, and 31 respectively in 2012. However, the number of mature fruits were relatively low (Table 18.3) compared to very highly significant ($P = 0.001$) tomato fruit yield of the genotypes planted in June (Table 18.1) in spite of NPK fertilizer use in the late planted tomato plants. Growing tomato late in the dry season exposes the crop to red spider mites (*Tetranychus evansi* Baker and Prichard) infestation resulting in highly significant ($P = 0.01$) tomato fruit loss. The loss of tomato fruits is caused by red spider mites infecting leaves causing the leaves to dry and senes. Highly significant ($P = 0.01$) tomato fruit losses caused by red mites were noted in the late planted tomato crop (Table 18.5).

Table 18.5 Fruit loss caused by red spider mites infestation in Tabora, Tanzania

Tomato genotype	Premature tomato fruits
Oxyl	23.1
Tanya Kibo	28.6
Tengeru	7.0
LSD 0.05	3.9
CV%	28.5

Conclusions

Currently, there is no information regarding tomato genotypes performance under elevated air temperatures, high evaporation and low atmospheric humidity. This study shows that application of balanced NPK inorganic fertilizers significantly improved tomato yield. The tomato genotypes Tanya Mkulima, Tanya Kibo and Oxyl performed better than Tangeru. Planting tomato late in August results in very low yield. More studies are needed to evaluate tomato genotypes tolerant to elevated dry season air temperatures in the semi-arid environments in Tanzania.

Acknowledgments The author wish to thank the farmers whose contribution and information was vital for carrying out field works. The publication of this paper has been possible by the support of the technical committee of the 27th SSSEA conference, in Nakuru, Kenya who selected this paper and made it possible to be spotted by Professor Walter Leal.

References

Abdul-Baki AA (1991) Tolerance of tomato cultivars and selected germplasm to heat stress. J Am Soc Horti Sci 116(6):1113–1116

Bagarama FM (2013) Red spider mites (*Tetranychus evansi*) in dry season agriculture in Tanzania. Paper presented at the 2nd international interdisciplinary annual conference,theme:the challenges of development in Africa, The Catholic University of Eastern Africa(CUEA), 25–29 June 2013

Boyer J (1982) Plant productivity and the Environment. Science 218(4571):443–448

Fischer G, Shah M, Veldhuisen HV (2002) Climate change and agricultural vulnerability. International Institute for Applied Systems Analysis. Report prepared under UN Institutional Contract Agreement 1113 for World Summiy on Sustainable Development.Laxenburg.Austria

Higashide T (2009) Prediction of tomato yield on the basis of solar radiation before anthesis under warm greenhouse conditions. HortScience 44:1847–1878

Ikilulu M (2006) Impacts of small reservoirs on rural livelihoods in Northern Ghana. M.Phil. Dissertation submitted to the Department of Agricultural Economics and Agribusiness, University of Ghana, Legon

Landon JR (1984) Booker tropical soil manual: a handbook for soil survey and agricultural land evaluation in the Tropics and Subtropics. Booker Agriculture Ltd, London, p 450

Lobell DB, Asner GP (2003) Climate and management contributions to recent trends in US agricultural yields. Science 299:1032

Lovatt J, Fullelove G, Wright R, Meurant N, Barnes J, O'Brien R (1998) Tomato Information Kit. Department of Primary Industries, Brisbane

Mongi H, Majule AE, Lyimo JG (2010) Vulnerability and adaptation of rain fed agriculture to climate change and variability in semi-arid Tanzania. Afr J Environ Sci Technol 4:371–381

Montterroso VA, Wien HC (1990) Flower and pod abscission due to heat stress in beans. J Am Soc Horti Sci 115:631–634

Nakuja T, Sarpong DB, Kuwornu JKM, Felix AA (2012) Water storage for dry season vegetable farming as an adaptive to climate change in the upper east region of Ghana. Afr J Agric Res 7:298–306

Ojo OD, Connaughton M, Kintomo AA, Olajide-Taiwo LO, Afolayan SO (2011) Assessment of irrigation systems for dry season vegetable production in urban and peri-urban zones of Ibadan and Lagos, Southwestern Nigeria. Afr J Agric Res 6(2):236–243

Ormrod DP, Woolley GW, Eaton A, Stobbe EH (1967) Effect of temperature on embryo sac development in *Phaseolus vulgaris L.* Can J Bot 45:948–950

Peet M, Willits D, Gardner R (1997) Response of ovule development and post-pollen production processes in male-sterile tomatoes to chronic, sub-acute high temperature stress. J Exp Bot 48:101–111

Sato S, Kamiyama M, Iwata T, Makita N, Furukawa H, Ikeda H (2006) Moderate increase of mean daily temperature adversely affects fruit set of *Lycopersicon esculentum* by disrupting specific physiological processes in male reproductive development. Ann Bot 97:731–738

Ssekabembe CK, Bukenya C, Nakyagaba W (2003) Traditional knowledge and practices in local vegetable production in central Uganda. Afr Crop Sci Conf Proc 6:14–19

Tsoho BA, Salau SA (2012) Profitability and constraints to dry season vegetable production under fadama in Sudan savannah ecological zone of Sokoto State, Nigeria. J Dev Agric Econ 4:217–222

Chapter 19
Drought Mitigating Technologies: An Overview of Cassava and Sweetpotato Production in Mukuyuni Division Makueni District in Semi-Arid Eastern Kenya

Cyrus M. Githunguri and Ruth L. Amata

Abstract Farmers in Mukuyuni can easily adopt drought mitigating technologies like cassava and sweetpotato. They are mainly propagated through stem cuttings or vines to produce starchy tuberous roots, which would provide the much needed carbohydrates if promoted properly. The results showed that about 90 % of the farmers put between 0.125 and 0.25 acres under cassava cultivation. The number of years under cassava and sweetpotato production ranged between 1 and 20 years with only a few farmers indicating to have been growing cassava for a period above 20 years. The majority of farmers were growing local cultivars of the two crops. The main method of utilizing cassava was boiling and eating as a snack (45 %). Over 77.8 % of the respondents indicated the origin of their cassava and sweetpotato cultivars as other farmers. The main method of utilizing sweetpotato was boiling and eating as a snack (56.3 %). About 3.1 % of the farmers mixed sweetpotato with beans and maize and consumed it as a stew while another 3.1 % have not utilized sweetpotato at all. Only 3.1 % fed sweetpotato to livestock. The rest, 28.1 % sold sweetpotato in the local market. This study established that there is a lot of room for commercializing cassava and sweetpotato production in Mukuyuni Division Makueni County through processing and promoting their utilization as a meal.

Keywords Cassava · Sweetpotato · Drought tolerant · Mitigation · Semi-arid areas

C.M. Githunguri (✉)
KARI Katumani Research Centre, P.O. Box 340-90100, Machakos, Kenya
e-mail: cyrusgithunguri@yahoo.com

R.L. Amata
KARI Kabete Research Centre, P.O. Box 147333–00800, Nairobi, Kenya

W. Leal Filho et al. (eds.), *Adapting African Agriculture to Climate Change*,
Climate Change Management, DOI 10.1007/978-3-319-13000-2_19

Introduction

Cassava and sweetpotato are drought tolerant crops, which are ideal for the arid, and semi-aridlands in provision of much needed calories (Githunguri et al. 2006). A lot of work has been carried out on processing, value addition, and utilization of these crops within KARI and other institutions (Nweke et al. 2002; Makokha and Tunje 2000; Wambugu and Mungai 2000; Githunguri 1995). It is very important to present these crops to urban consumers in an attractive form at affordable prices, which are competitive to those of cereals (Nweke et al. 2002). Cassava and sweetpotato products processing and utilization is done mainly at the subsistence level (Kadere 2002) and in Kenya is still limited to the household level. Past studies show that about 80 % of the cassava and sweetpotato products are consumed on the farm while 20 % are marketed (CBS 1998). At the household level cassava is mainly utilized as u*gali*, which is prepared from cassava composite flour (Githunguri 1995). Some development agencies, KARI, and the Home Economics branch of Ministry of Agriculture have been promoting different recipes from cassava and sweetpotato flours especially baked products such as cakes (Githunguri et al. 2006). Cassava and sweetpotato production in Kenya is constrained by lack of adequate disease and pest free planting materials due to their slow multiplication rate and poor cultural practices among other factors (Githunguri et al. 2003; Odendo et al. 2001). There has been more emphasis on cultivar development than on agronomic packages and as such, there is urgent need for research programmes to start addressing agronomic requirements and rapid multiplication and distribution of clean elite planting material. The objectives of the Root and Tuber Crops Programme in Katumani are to develop cassava and sweetpotato varieties that are widely adapted to diverse agro-ecological zones. The varieties should also be high yielding, early bulking, and drought resistant/tolerant, resistant to major biotic and abiotic stresses and have good root quality (Githunguri et al. 2003; Githunguri 2004). KARI-Katumani has recognized the importance of involving farmers in their selection and breeding research programmes as suggested by Bellon (2001) and Fliert and Braun (1999). The goal of the project is to promote and distribute early maturing, drought, disease, and pest tolerant cassava and sweetpotato cultivars available at KARI Katumani in the semi-arid areas of Kenya starting with the selected project sites. The specific objective of the study was to obtain a general overview of cassava and sweetpotato production in Matiliku Division Makueni District in Semi-Arid Eastern Kenya as a prelude to the establishment of a seed system for the two crops. This will ensure the elite KARI bred high yielding cassava and sweetpotato varieties will move out and have the desired impact on food security and improved livelihoods in semi-arid eastern Kenya.

Materials and Methods

Participatory rural appraisal (PRA) interviews took place at a farm in Mukuyuni located at Latitude South: 01.73256 and Longitude East: 037.42758 with an altitude of 1,366 m above sea level on 11th June 2010 involving 13 female and 7 male farmers. KARI and extension officers conducted the PRA. During the exercise, the status of cassava and sweetpotato production in Mukuyuni was established. The results of PRA were analyzed using descriptive statistics.

Results and Discussion

Figure 19.1 shows about 90 % of the farmers put between 0.125 and 0.25 acres under cassava cultivation. The rest of the farmers put in only a few stands of cassava in their farms. According to Fig. 19.2, the number of years under cassava production ranged between 1 and 20 years with only a few farmers indicating to have been growing cassava for a period above 20 years. About 45 % of the farmers have grown cassava for about 15 years while the rest had cassava-growing experience ranging between 1 and 10 years in proportions ranging between 5.6 and 11.1 %. These results suggest that cassava is still being grown as a subsistence crop and a lot needs to be done if cassava is to be commercialized in Mukuyuni.

Figure 19.3 shows 67 % of the farmers were growing a local cultivar known as Kitwa followed by Mulava at 17 %. The proportion of farmers growing the rest of the cultivars Kilava, Meu and an unknown improved cultivar ranged between 4 and 8 %. This suggests that the adoption of improved cassava cultivars has been slow and as such there is need to put a lot of effort in the promotion of improved cultivars in Mukuyuni.

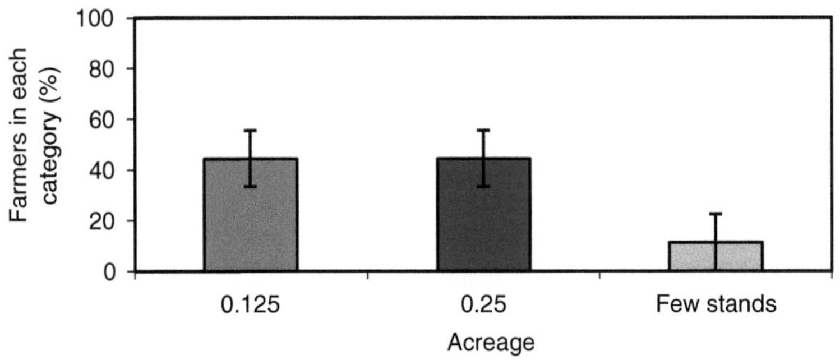

Fig. 19.1 Proportion of acreage under cassava production in Mukuyuni division Makueni district during 2010 long rains cropping season. *Vertical bars* represent the standard error between means ($P = 0.05$)

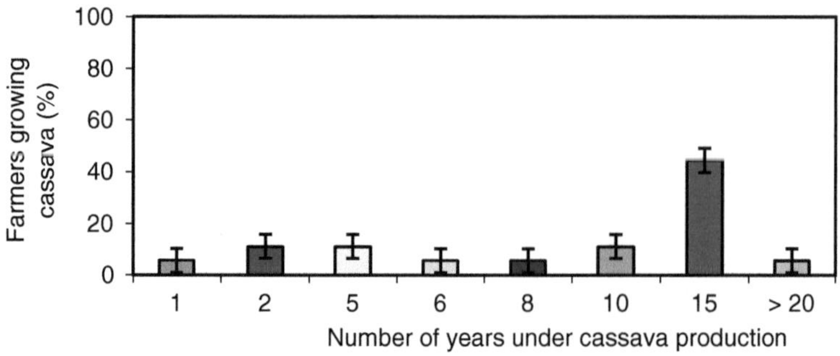

Fig. 19.2 Number of years under cassava production and proportion of farmers growing them in Mukuyuni division Makueni district. *Vertical bars* represent the standard error between means ($P = 0.05$)

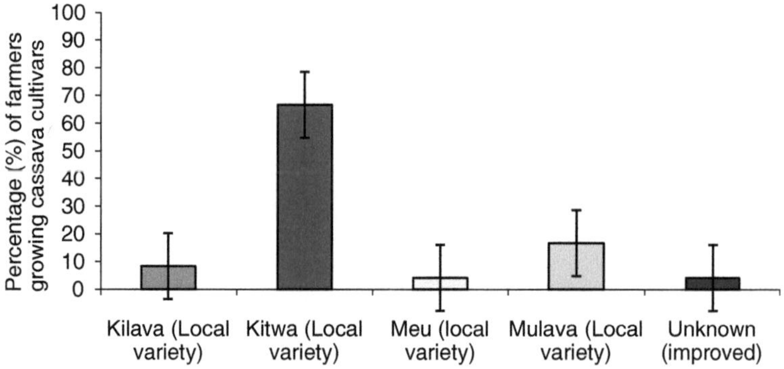

Fig. 19.3 Types of cassava cultivars cited and proportion of farmers growing them in Mukuyuni division Makueni district during 2010 long rains cropping season. *Vertical bars* represent the standard error between means ($P = 0.05$)

Table 19.1 shows that the main method of utilizing cassava was boiling and eating as a snack (45 %). 12.5 % of the farmers mixed cassava with beans and maize and consumed it as a stew while 10 % chewed cassava raw. Very few farmers, about 7.5 %, mixed cassava either with maize, millet, or sorghum to form composite flour for "ugali". Only 2.5 % fed cassava to poultry. The rest, 22.5 % sold cassava in the local marketing. This means that very few farmers process cassava or consume cassava as a major part of their diet. There is a lot of room for commercializing cassava and promoting its utilization as a meal.

During the PRA 77.8 % of the respondents indicated the origin of their cassava cultivars as other farmers, 5.6 % as KARI and a similar 5.6 % as coast, while 11.1 % did not know their origin. On the other hand, 80 % of the respondents indicated the origin of their sweetpotato cultivars as other farmers, 10 % as KARI, while 10 % did not know their origin.

Table 19.1 Method of utilization of cassava and proportion of farmers involved in Mukuyuni division Makueni district as of 2010 long rains cropping season

Method of utilization	Percentage (%) of farmers utilizing them
Boil as a snack	45
Chew raw	10
Feed poultry	2.5
Mill into composite flour (maize and cassava) for "ugali"	2.5
Mill into composite flour (millet and cassava) for "ugali"	2.5
Mill into composite flour (millet or sorghum and cassava) for "ugali"	2.5
Mix with beans and maize	12.5
Sell in the market	22.5

Figure 19.4 shows 36.8, 42.1, 10.5, and 5.3 % proportion of respondents grow sweetpotato 0.125, 0.25, 0.5, 0.75 acres, and a few stands, respectively. These results suggest that like cassava, sweetpotato is still being grown as a subsistence crop and a lot needs to be done in order to commercialize it in Mukuyuni. However, it seems the area under sweetpotato is bigger than the area under cassava.

According to Fig. 19.5, the number of years under sweetpotato production ranged between 1 and 20 years with only very few farmers indicating to have been growing sweetpotato for a period above 20 years. About 42 % of the farmers have grown sweetpotato for about 15 years while the rest had experience on sweetpotato production ranging between 1 and 17 years in proportions ranging between 5.3 and 21.1 %. Similar to cassava, these results suggest that sweetpotato is also largely being grown as a subsistence crop and a lot of efforts needs to be taken in order to commercialize sweetpotato in Mukuyuni. Such efforts include the establishment of a sustainable vigorous cassava and sweetpotato seed system in place.

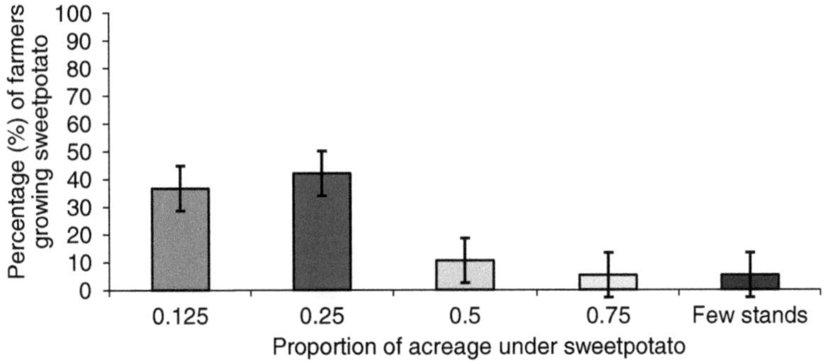

Fig. 19.4 Proportion of acreage under sweetpotato production in Mukuyuni division Makueni district during 2010 long rains cropping season. *Vertical bars* represent the standard error between means ($P = 0.05$)

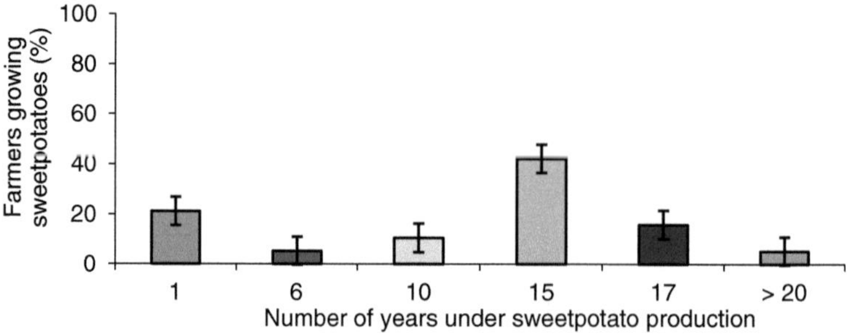

Fig. 19.5 Number of years under sweetpotato production and proportion of farmers growing them in Mukuyuni division Makueni district. *Vertical bars* represent the standard error between means ($P = 0.05$)

According to Fig. 19.6, 33.3, 4.2, 25.0, 4.2, 8.3, 8.3, 8.3, and 8.3 %, farmers were growing sweetpotato cultivars Kiluu (Local variety), Kitune (Local variety), local variety (unknown), Meu (Local variety), Mutune (Local variety), Mwezi moja (Local variety), Orange fleshed, and an Unknown (improved). This suggests that the adoption of improved sweetpotato cultivars has been slow and as such there is need to put a lot of effort in the promotion of improved cultivars in Mukuyuni. However, it seems the farmers were keen on sourcing more sweetpotato cultivars than cassava.

Table 19.2 shows that the main method of utilizing sweetpotato was boiling and eating as a snack (56.3 %). 3.1 % of the farmers mixed sweetpotato with beans and maize and consumed it as a stew while another 3.1 % have not utilized sweetpotato at all. Very few farmers (6.3 %) roasted sweetpotato and ate them as a snack. Only 3.1 % fed sweetpotato to livestock. The rest, 28.1 % sold sweetpotato in the local

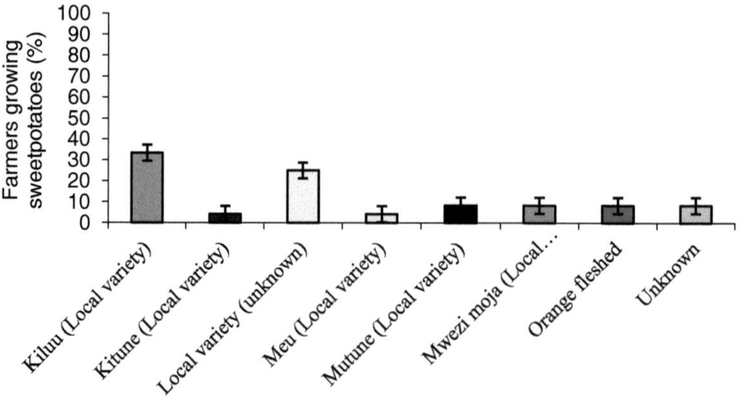

Fig. 19.6 Types of sweetpotato cultivars cited and proportion of farmers growing them in Mukuyuni division Makueni district during 2010 long rains cropping season. *Vertical bars* represent the standard error between means ($P = 0.05$)

Table 19.2 Method of utilization of sweetpotato and proportion of farmers involved in Mukuyuni division Makueni district as of 2010 long rains cropping season

Method of utilization	Percentage (%) of farmers utilizing them
Boil as a snack	56.3
Feed livestock	3.1
Have not utilized	3.1
Mix with beans and maize	3.1
Roast	6.3
Sell in the market	28.1

market. This means that very few farmers consume sweetpotato as a major part of their diet. There is a lot of room for commercializing sweetpotato through processing and promoting their utilization as a meal.

Table 19.3 shows the education level, their occupation, and main mode of communication by farmers involved in cassava and sweetpotato production in Mukuyuni. The majority of farmers, 38.9 %, had attained the Primary School level of education, were farmers, and owned a cellphone. Eleven percent of the respondents had not attained Primary School level of education (Adult Education) and did not own a cellphone. A similar percentage of respondents who had either attained the Primary School level or Secondary School level of education did not own a cellphone. This suggests that there is no clearly defined relationship between ownership of a cellphone and level of education among the respondents. However, it seems the majority of those who owned a cellphone had attained the Primary School level of education. Ownership of a cellphone is an indication of farmers who can easily adopt a new technology if promoted properly.

Table 19.3 Education level, occupation and main mode of communication by farmers involved in cassava and sweetpotato production in Mukuyuni division Makueni district during 2010 long rains cropping season

Education level	Occupation	Telephone	Farmers within the main categories (%)
Adult education	Farmer	None	11.1
Primary	Church leader/farmer	Cellphone	5.6
Primary	Farmer	Cellphone	38.9
Primary	Farmer	None	11.1
Primary	Mason/farmer	Cellphone	5.6
Secondary	Farmer	Cellphone	5.6
Secondary	Farmer	None	11.1
Secondary	Mason/farmer	Cellphone	5.6
Secondary	Nursery teacher/farmer	Cellphone	5.6

Conclusions and Recommendations

The farmers' cassava and sweetpotato growing experience is short. The area under sweetpotato is bigger than the area under cassava and it seems the farmers were keen on sourcing more sweetpotato cultivars than cassava. These crops are still being grown as subsistence crops and a lot needs to be done if they are going to be commercialized in Mukuyuni. Use of improved cassava and sweetpotato cultivars from research institutions was very low. The adoption of improved cassava and sweetpotato cultivars has been slow and as such there is need to put a lot of effort in the promotion of improved cultivars in Mukuyuni. Such efforts include the establishment of a sustainable vigorous cassava and sweetpotato seed system in place. In addition, very few farmers process or consume cassava and sweetpotato as a major part of their diet. There is a lot of room for commercializing these crops through processing and promoting their utilization as a meal. Farmers in Mukuyuni can easily adopt drought mitigating technologies if promoted properly.

Acknowledgment The invaluable logistical and financial resources that made this work possible provided by the Director KARI, and USAID are gratefully acknowledged. The important technical contributions by the Root and Tuber Crops Programme staff are also acknowledged with gratitude.

References

Bellon MR (2001) Participatory research methods for technology evaluation: a manual for working with farmers. CIMMYT, Mexico, 93 p. Central Bureau of Statistics. 1998 Statistical Abstract, Nairobi, Kenya

Githunguri CM (1995) Cassava food processing and utilization in Kenya. In: Egbe TA, Brauman A, Griffon D, Treche S (eds) Cassava food processing. CTA, ORSTOM, pp 119–132

Githunguri CM (2004) Farmers' participatory perspectives on sweetpotato cultivars in Kathiani division of Machakos district, Kenya. In: Book of abstracts of the 9th triennial symposium of the international society for tropical root crops—Africa Branch (ISTRC-AB), Mombasa, Kenya, 31 Oct–5 Nov 2004, 84 p

Githunguri CM, Migwa YN, Ragwa SM, Karoki MM (2003) Cassava and sweetpotato agronomy, physiology, breeding, plant protection and product development. Root and tuber crops programme in KARI-Katumani. Paper presented at the joint planning meeting organized under the eastern province horticulture and traditional food crops project, held at Machakos, Kenya, on 5, 7 Mar 2003, 5 p

Githunguri CM, Karuri EG, Kinama JM, Omolo OS, Mburu JN, Ngunjiri PW, Ragwa SM, Kimani SK, Mkabili DM (2006) Sustainable productivity of the cassava value chain: an emphasis on challenges and opportunities in processing and marketing Cassava in Kenya and beyond. A project proposal presented to the Kenya agricultural productivity project (KAPP) competitive agricultural research grant fund. Research call ref no. KAPP05/PRC-CLFFPS-03. KAPP Secretariat, 106 p

Kadere TT (2002) Marketing opportunities and quality requirements for cassava starch in Kenya. In: Proceedings of regional workshop on improving the cassava sub-sector, held in Nairobi, Kenya, 8–18 Apr 2002, pp 81–86

Makokha J, Tunje TK (2000) Study of traditional utilization and processing of cassava and cassava products in Kenya. First interim technical and financial report, JKUAT/EARRNET

Nweke FI, Spencer DSC, Lynam JK (2002) Cassava transformation. International Institute of Tropical Agriculture, Nigeria, 273 p

Odendo M, kilambya DW, Ndolo PJ (2001) Sweetpotato research priorities at the Kenya agricultural research institute. Paper presented at the stakeholders workshop for sweetpotato priority setting, held at Machakos, Kenya, 27–28 Sept 2001, 19 p

van de Fliert E, Braun AR (1999) Farmer field school for integrated crop management of sweetpotato: field guides and technical manual. Andi Offset, CIP, Yogyakarta, Indonesia, III-101 p

Wambugu SM, Mungai JN (2000) The potential of cassava as an industrial/commercial crop for improved food security, employment generation and poverty reduction in Kenya. Kenya Industrial Research and Development Institute (KIRDI), Kenya

Chapter 20
Cassava Farming Transforming Livelihoods Among Smallholder Farmers in Mutomo a Semi-arid District in Kenya

Cyrus M. Githunguri, Esther G. Lung'ahi, Joan Kabugu and Rhoda Musili

Abstract The study established that climate change is real and has negatively affected smallholder farmer families in Mutomo and as such it was prudent to introduce drought tolerant crops like cassava in order to improve food security as a climate change adaptation technology. It was evident that the elite cassava varieties from KARI were supplying the much needed carbohydrates in an affordable form. The assessment also established water scarcity is a major development-limiting factor in Mutomo that needed urgent attention. The Mutomo community had shed "*Mwolyo*" the hand-out mentality through adoption of appropriate technologies for this place like growing, processing, marketing, and consumption of cassava. Cassava roots were mainly marketed as fresh roots for chewing and boiling. Cassava cuttings and cakes on sale in the Mutomo market suggested that demand for cassava was rising and only need upscaling. Being dominated by agro-pastoralists, it was obvious that cassava and other crops that take more than four months before being harvested do not fit well into the system and this is an area that has to be addressed to pre-empt potential conflicts.

Keywords Cassava · Climate change · Adaptation · Semi-arid areas

C.M. Githunguri (✉)
KARI Katumani Research Centre, P.O. Box 340-90100, Machakos, Kenya
e-mail: cyrusgithunguri@yahoo.com

E.G. Lung'ahi · J. Kabugu
Arid Lands Information Network (ALIN), State House Crescent, 10098 00100, Nairobi, Kenya

R. Musili
Arid Lands Information Network (ALIN), Mutomo Maarifa Centre, Mutomo, Kitui, Kenya

W. Leal Filho et al. (eds.), *Adapting African Agriculture to Climate Change*, Climate Change Management, DOI 10.1007/978-3-319-13000-2_20

Introduction

Cassava (*Manihot esculenta* Crantz) produces about 10 times more carbohydrates than most cereals per unit area, and are ideal for production in marginal and drought prone areas, which comprise over 80 % of Kenya's land mass (Githunguri et al. 1998; Githunguri 2002; Nweke et al. 2002). A cassava plant possesses several growth parameters and physiological processes which can be used to measure its ability to produce adequate yield under various abiotic and biotic stresses (Ekanayake et al. 1997a, b; Ekanayake 1998; IITA 1982, 1990a; Osiru et al. 1995). According to these authors, some of these parameters include long fibrous roots, shedding of leaves, leaf area index, leaf water potential, moderate stomatal conductance, transpiration rate, water use efficiency, crop growth rate and dry matter accumulation in the tuberous roots. Cassava can reach its production potential only where the attributes of the environment best match the crop requirements. Breeding and selection of varieties according to prevailing environmental characteristics can ensure optimal performance (IITA 1990b).

The cassava commodity system has four main components: production, processing, marketing and consumption. Linking them is the key to successful cassava products development. Strong ties with both public and private institutions engaged in research, extension and social development are essential in the accomplishment of this linkage. The exact character of these linkages will vary according to the stage of the project in technology generation and transfer (Githunguri et al. 2006). Plant breeders can contribute to better productivity and quality, agronomists to improvements in cultural practices and cropping systems, and agro-ecologists to the proper analysis of resource management issues. In order to enhance the commercial achievement of Economic Recovery Strategy (ERS) goals, the Government of Kenya in collaboration with development partners has established the Kenya Agricultural Productivity Project (KAPP) (Ministry of Agriculture 2005). The KAPP uses thematic concepts with demonstrative multi-sectoral approach to address agricultural challenges. In this regard, the cassava value chain project was funded by KAPP to enhance cassava production, processing and marketing in Kenya and beyond our borders, especially the Common Market for Eastern and Southern Africa (COMESA) region and Europe (Kadere 2002; Mbwika 2002). In Eastern Kenya cassava is eaten either raw or boiled (Githunguri 1995). Despite its great potential as a food security and income-generating crop among rural poor in marginal lands, its utilization remains low. The potential to increase its utilization is enormous with increased recipe range (Githunguri 1995) and provision of adequate clean planting material. One of the major constraints to cassava production in the arid and semi-arid areas includes lack of adequate disease and pest free planting materials (Obukosia et al. 1993) exacerbated by the slow multiplication rates of 1:10. KARI-Katumani has bred cultivars tolerant to cassava mosaic disease and acceptable to end-users (Githunguri et al. 2003) whose multiplication and distribution is being attained through irrigation. Other constraints to cassava production in Kenya including semiarid eastern include lack of adequate disease and pest free planting materials, poor cultural practices, lack of appropriate storage and processing technologies, poor market infrastructure (Githunguri and Migwa 2003; Lusweti et al. 1997). KARI-Katumani has developed cassava varieties that are widely adapted to diverse agro-ecological zones, high yielding, early bulking, drought resistant/tolerant, resistant to major biotic and abiotic stresses and have good root quality (Githunguri et al. 2003; Githunguri 2004). KARI-Katumani has recognized the importance of involving farmers in their selection and breeding research programmes as suggested by Bellon (2001) and Fliert and Braun (1999). One of the objectives of this project was to establish demonstration plots and select entrepreneurial farmers for commercial cassava planting material multiplication and distribution to farmers in semi-arid Eastern Kenya in a bid to improve food security as a climate change adaptation technology. Mutomo district situated in Kitui County in semi-arid Eastern Kenya was one of the areas that were selected for the establishment of a cassava seed system.

Materials and Methods

Cassava agronomic demonstrations, seed multiplication and distribution programs were established in order to assure processors of a steady supply of tuberous roots. To ensure sustainable supply of planting materials, farmers who were willing to grow cassava on at least a quarter of an acre of their farm were selected to participate in planting materials multiplication and distribution in the selected project sites, Kibwezi (Kwa-Kyai), Mukuyuni, Mutituni, Matiliku, and Mutomo. Elite cassava cultivars grown under sprinkler irrigation at KARI Masongaleni and Kiboko Sub-Centers situated in Kibwezi and Makindu districts were harvested and distributed to several farmers who had shown interest in growing cassava in the project sites. In 2009, Revitalization of Indigenous Initiatives for Community Development (RINCOD), a local non-governmental organization carried out needs assessments in Mutomo. During these assessments, community groups agreed to start the Mutomo Cassava Production and Processing Association (MUKAPA). The project involved more than 100 households and over time, it has established high-yielding, disease and drought-resistant cassava varieties developed by the Kenya Agricultural Research Institute (KARI). During project initiation, the community selected three farms for cassava propagation and planted 10,000 cuttings on each with support from RINCOD and KARI. From these three farms, all the 100 members were then supplied with cuttings. During the last quarter of year 2012 and 1st quarter of 2013, A team of journalists, extension and research officers went out on a fact finding tour to assess the impact the introduction of elite cassava had on the livelihoods of participating farmers in Mutomo using photography and focussed group discussions.

Results and Discussions

The use of the elite cassava varieties has had several benefits among the Mutomo community. Figure 20.1 shows the Mutomo Meteorological Officer collecting dairy weather data. According to the officer, they are mainly concerned with temperature, armyworm, and rainfall data. Rainfall in Mutomo is scarce and erratic, while temperatures are generally high and the area prone to armyworm infestations especially during good seasons and that is why it is crucial to monitor these parameters.

A team of journalists, extension and research officers visited a typical Mutomo farmer's homestead to get a first-hand experience of what he/she goes through in a normal day (Fig. 20.2). This farmer like most others has been growing maize and beans religiously season after season despite frequent crop failures. Changing the farmer's attitude towards the growing of these crops is a major challenge in tackling food insecurity in this area. Like the pictures depict, poverty levels and food insecurity are very high. The empty granary is a constant and stark reminder of the frequent crop failures that have occurred in that area since year 2003. The owner of this homestead works as a night guard in Mutomo town and as such he had been able to install a small solar panel that is able to light four bulbs and run a small radio. Learnt climate change is real and has negatively affected Mutunga's family. It was sad to see the farmer's abandoned empty granary due to climate change since year 2003. The team realized how lucky the farmers in high potential areas are and the urgency there is in creating awareness among them about the utter need for them

Fig. 20.1 Mutomo meteorological officer collecting (i) max. temperature, (ii) armyworm, and (iii) rainfall data

Fig. 20.2 Mutunga's homestead (i and ii) and a (iii and iv) granary which has been empty for several failed seasons since year 2003–this is typical of a smallholder farmer in Mutomo who has been growing maize and beans

to preserve, conserve and improve their environment jealously. There is real need to introduce drought tolerant crops like cassava in this area if food insecurity is going to be addressed.

Figure 20.3 shows the beginning of a typical day of a smallholder farmer in Mutomo. The farmer and his family which included the family cat were enjoying boiled cassava for breakfast. Due to the high poverty levels in the area the most of the farmers cannot afford to buy bread for breakfast. The family heartily shared the little cassava- breakfast which was really moving. The pictures suggest cassava is important in supplying the much needed carbohydrates in an affordable form. However, since this is a predominantly "boil and eat" society, it is important to introduce only low cyanogenic cassava cultivars. In addition, it is crucial to train cassava consumers on how to detoxify cassava through appropriate processing.

After the farmer had taken cassava breakfast with his family the team walked to his main farm which was about 20 min away. During the walk the team of officers was able to observe other resident farmers going to fetch water in a common well situated near the farmer's main farm (Fig. 20.4). Looking at the heavy traffic consisting of both livestock and human beings going to collect water at the same well, it was realized that water scarcity is a major development-limiting factor in this area that needs urgent attention. Figure 20.5 shows Mutunga's daughter in-law and other neighbours fetching water from a dry riverbed well situated 3 km away from their homestead. In this area the donkey is the preferred mode of transporting water from the communal wells. It was heartrending to watch a whole population including men, women, and children, and their livestock going for water in just one watering place where there was not enough of it. However, it was encoufinger milletng to realize that the community here was doing all it can to make ends meet and that they had shed "*Mwolyo*" the handout mentality through adoption of

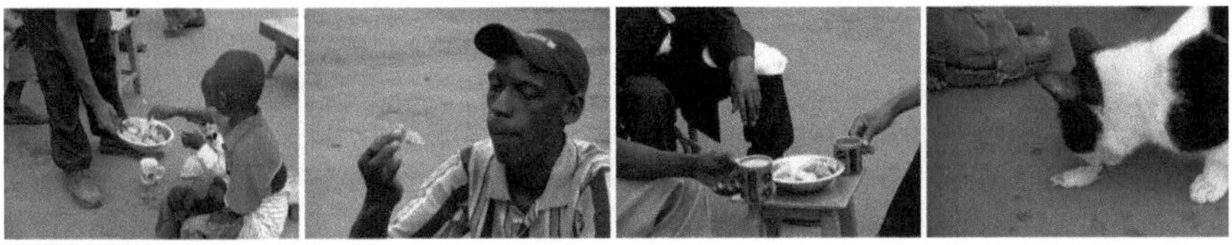

Fig. 20.3 Mutunga enjoying cassava breakfast together with his grandchildren, son and family cat enjoying

Fig. 20.4 Smallholder farmers fetching water in a communal well situated in a dry riverbed over 3 km away from most homesteads

Fig. 20.5 Children and women scoop water from a typical shallow well in Kanzilu location in Mutomo for domestic use and for their donkeys and other livestock

appropriate technologies for this place like growing, processing, marketing, and consumption of cassava.

Water scarcity in Kanzilu is a major challenge as is clearly evident from the pictures in Fig. 20.5. Children and women scoop water from a typical shallow well in Kanzilu location in Mutomo for domestic use and for their donkeys, cows and goats. As is also evident getting clean water is also a major challenge. There is need to introduce project s that supply adequate clean water in this area alongside those addressing food insecurity. Transporting water and watering livestock in Kanzilu is a major occupation mainly conducted by women and children.

Water scarcity in Kanzilu, a village in Mutomo, is a major challenge in the lives of residents and has to be addressed through diverse approaches. According to Fig. 20.6, cassava is a mitigating technology. The farmer's cassava farm had been maliciously grazed by goats which unless addressed is a major potential area for conflict between farmers and agro-pastoralists. It was difficult to understand why anybody in their right frame of mind would want to deliberately graze his goats on the farmer's farm. However, on the flipside this demonstrates the importance and resilience of cassava as a drought mitigating technology in Mutomo. The farming system in Mutomo is dominated by agro-pastoralists and it is apparent that cassava and other crops that take more than 4 months before being harvested do not fit into the system as farmers release their livestock to graze communally after maize has been harvested. This is an area that has to be addressed if cassava farming is going to be adopted. However, the farmers understand this and they fence the parts of their farms that have crops that take longer than 4 months to mature.

According to Fig. 20.7, cassava was mainly marketed as fresh roots for chewing by men to improve their virility. Cassava cuttings were also on sale in the Mutomo market suggesting demand for cassava was rising. Cassava cakes were also on

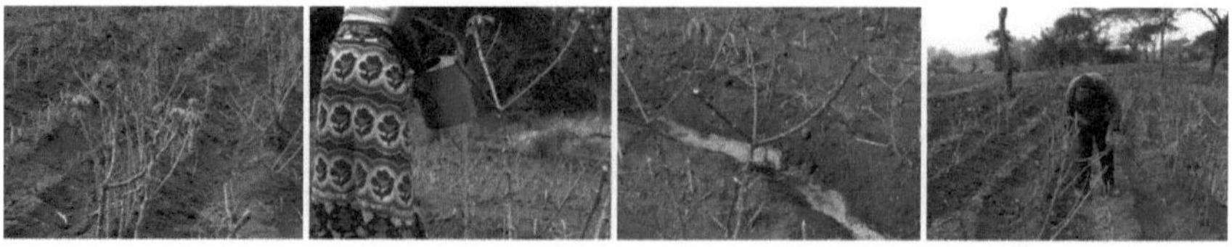

Fig. 20.6 Watering cassava in Mutunga's farm in Kanzilu village in Mutomo which had been grazed and defoliated by goats maliciously belonging to agro-pastoralists

Fig. 20.7 A typical day in Mutomo market: (i) a vendor selling cassava & assorted fruits; (ii) cassava roots for chewing on sale; (iii) cassava stakes on sale; and (iv) cassava-based baked products on sale

Fig. 20.8 (i) Cassava cake ingredients; (ii) cassava cake dough being spread in the pot ready for baking; (iii) and (iv) cassava cake is finally ready for serving; and (v) Mutheu can hardly wait to taste the cassava cake

display in the market. The photographs suggest that cassava products were popular in Mutomo and that is why they were on sale. Cassava roots, cuttings, and cakes were on sale in the open-air market.

At the Mutomo Bakery, cassava was being chipped, dried in a solar drier, milled and bread baked using the same flour mixed with wheat flour in various ratios. A loaf of cassava bread was retailing at K. Shs 42. Figure 20.8 shows a demonstration on how to make a cassava-based cake which has a ready market in Mutomo.

Conclusions and Recommendations

Climate change is real and has negatively affected smallholder farmer families in Mutomo. There is urgent need to introduce drought tolerant crops like cassava in Mutomo in order to address food insecurity. It was evident the elite cassava varieties from KARI have brought several benefits among the Mutomo community. Due to the high poverty levels in the area the most of the farmers cannot afford to buy bread for breakfast and as such cassava is important in supplying the much needed carbohydrates in an affordable form. Water scarcity is a major development-limiting factor in Mutomo that needs urgent attention. As such there is need to introduce projects that supply adequate clean water in this area alongside those addressing food insecurity. The Mutomo community was doing all it can to make ends meet and that they had shed "*Mwolyo*" the handout mentality through adoption of appropriate technologies for this place like growing, processing, marketing, and consumption of cassava. Cassava roots were mainly marketed as fresh roots for chewing by men to improve their virility. Cassava cuttings were also on sale in the Mutomo market suggesting demand for cassava was rising. Cassava cakes were also on display in the market suggesting that cassava products were popular in Mutomo and only need upscaling. The farming system in Mutomo is dominated by agro-pastoralists and it is apparent that cassava and other crops that take more than 4 months before being harvested do not fit well into the system as farmers release their livestock to graze communally after maize has been harvested. This is an area that has to be addressed to pre-empt potential conflicts. However, the farmers understand this and they fence the parts of their farms that have crops that take longer than 4 months to mature.

Acknowledgment The authors would like to acknowledge the Director KARI; the Director Arid lands Information Network, and the USAID Programme for providing funds, support and other facilities in making this work a success. The Director KARI is further thanked for granting permission to publish this work. We also thank Dr. A. M. Mailu and Ms. Jayne Gathii of USAID for many valuable discussions and suggestions. Finally but not last, we thank most sincerely the various collaborative partners (Ministry of Agriculture among others) for their invaluable technical contributions.

References

Bellon MR (2001) Participatory research methods for technology evaluation. A manual for working with farmers. CIMMYT, Mexico, D.F, p 93

Ekanayake IJ, Osiru DSO, Porto MCM (1997a) Morphology of cassava. IITA Res Guide 61:1–30

Ekanayake IJ, Osiru DSO, Porto MCM (1997b) Agronomy of cassava. IITA Res Guide 60:1–30

Ekanayake IJ (1998) Screening for abiotic stress resistance in root and tuber crops. IITA Res Guide 68:46

Githunguri CM, Migwa YN (2003) Sweetpotato feathery mottle virus resistance and yield characteristics of different Sweetpotato cultivars in Machakos and Makueni districts of Kenya. In: Githunguri CM, Kwena K, Kavoi J, Okwach EW, Gatheru M, Abok JO (eds) Kenya agricultural research institute, KARI Katumani research centre annual report, 2002, pp 91–95

Githunguri CM (1995) Cassava food processing and utilization in Kenya. In: Egbe TA, Brauman A, Griffon D, Treche S (eds) Cassava food processing. CTA, ORSTOM, Wageningen, pp 119–132

Githunguri CM, Ekanayake IJ, Chweya JA, Dixon AGO, Imungi JK (1998) The effect of different agroecological zones and plant age on the cyanogenic potential of six selected cassava clones. In: Ferris RSB (ed) Post harvest technology and commodity marketing. Proceedings of a postharvest conference held on 2 Nov–1 Dec 1995, Accra, Ghana. IITA, Ibadan, Nigeria, pp 71–76

Githunguri CM (2002) The influence of agro-ecological zones on growth, yield and accumulation of cyanogenic compounds in cassava. A thesis submitted in full fulfilment for the requirements for the degree of Doctor of Philosophy in Crop Physiology, Faculty of Agriculture, University of Nairobi, p 195

Githunguri CM, Migwa YN, Ragwa, SM, Karoki, MM (2003) Cassava and sweetpotato agronomy, physiology, breeding, plant protection and product development. Root and tuber crops programme in KARI-Katumani. Paper presented at the Joint Planning meeting organized under the eastern province horticulture and traditional food crops project, held at Machakos, Kenya on 5th–7th Mar 2003, p 5

Githunguri CM (2004) Farmers' participatory perspectives on sweetpotato cultivars in Kathiani division of Machakos district, Kenya. In: Book of abstracts of the 9th triennial symposium of the international society for tropical root crops-Africa branch (ISTRC-AB), Mombasa, Kenya, 31st Oct–5th Nov 2004, p 84

Githunguri CM, Karuri EG, Kinama JM, Omolo OS, Mburu JN, Ngunjiri PW, Ragwa SM, Kimani SK, Mkabili DM (2006) Sustainable productivity of the cassava value Chain: an emphasis on challenges and opportunities in processing and marketing cassava in Kenya and beyond. KAPP competitive agricultural research grant fund, p 106

IITA (1982) Management practices for production of cassava planting materials. IITA tuber and root crops production manual series, p 244

IITA (1990a) Cassava in tropical Africa. Reference manual IITA, p 176

IITA (1990b) Targeting cassava breeding and selection. In: Proceedings of the 4th West and Central Africa root crops workshop, held in Lome, Togo, 12–16 Dec 1988, IITA meeting reports series 1988/6, pp 27–30

Kadere TT (2002) Marketing opportunities and quality requirements for cassava starch in Kenya. In: Proceedings of regional workshop on improving the cassava sub-sector, held in Nairobi, Kenya, 8-18 Apr 2002, pp 81–86

Lusweti CM, Kiiya W, Kute C, Laboso A., Nkonge C, Wanjekeche E, Lobeta T., Laya S, Kakuko A, Chelang E (1997) The farming systems of Sebit. In: summary from PRA activities, pp 54–67

Ministry of Agriculture (MoA) (2005) Strategy for revitalizing agriculture 2004–2014. Ministry of agriculture, p 23

Mbwika JM (2002) Cassava sub-sector analysis in the Eastern and Central African region. In: Proceedings of regional workshop on improving the cassava sub-sector, held in Nairobi, Kenya, 8–18 Apr 2002, pp 8–18

Nweke FI, Spencer DSC, Lynam JK (2002) Cassava transformation. International Institute of Tropical Agriculture, Nigeria, p 273

Obukosia SD, Muriithi A, Musangi RS (1993) Biotechnological approach to the improvement of root, tuber and horticultural crops in Kenya. Production constraints and potential solutions. In: Proceedings of the national agricultural biotechnology workshop, Nairobi, pp 92–106

Osiru DSO, Porto MCM, Ekanayake IJ (1995) Physiology of cassava. IITA Res Guide 55:3–19

van de Fliert E, Braun AR (1999) Farmer field school for integrated crop management of sweet potato. Field Guides and Technical Manual. Andi Offset, CIP, Yogyakarta, Indonesia, pp III–101

MIX
Papier aus verantwortungsvollen Quellen
Paper from responsible sources
FSC® C105338

If you have any concerns about our products,
you can contact us on
ProductSafety@springernature.com

In case Publisher is established outside the EU,
the EU authorized representative is:
Springer Nature Customer Service Center GmbH
Europaplatz 3, 69115 Heidelberg, Germany

Printed by Libri Plureos GmbH
in Hamburg, Germany